# PRAISE FOR *DEATH MAKES LIFE POSSIBLE*

"Marilyn Schlitz's new book, *Death Makes Life Possible*, is a rare and wonderful exploration of that area of our culture that is still 'the most repressed topic': death. From a variety of perspectives—personal, spiritual, indigenous, traditional, post-materialist scientific—Dr. Schlitz opens this area to a fascinating adventure, including dialogues with dozens and dozens of teachers, scientists, spiritual leaders . . . all in an attempt to 'de-repress' this area and make it a part of the tapestry of life that it in fact is. Bold, brave, wise, riveting, her account superbly helps to overcome this 'denial of death' our culture is still saturated with."

KEN WILBER
Author of *The Integral Vision*

"This book courageously deals with the passage of death in both scientific and spiritual terms. Marilyn Schlitz is a superb teacher and scientist who has dedicated her life to compassionately enlightening people and stretching the bounds of limited scientific thinking to encompass the spiritual and intuitive. Highly recommended!"

JUDITH ORLOFF, MD
Author of *Second Sight*

"Helpful as well as delightful. With brilliant intellect and warm heart, Dr. Schlitz beautifully makes the age-old case that confrontation and contemplation of death brings one more vibrantly to life! I heartily recommend it."

ROBERT "TENZIN" THURMAN
Jey Tsong Khapa professor of Buddhist Studies,
Columbia University; founder, Menla Mountain Retreat

"We live in an exciting and edgy time when post-materialist scientists are incorporating some of the wisdom of ancient spiritual teachings while healthy spirituality eagerly interacts with science and together new and ancient questions are being explored freshly. This book, while incorporating science, art, dreams, and deep ecumenism, focuses on death and questions of life and/or consciousness after death. The author offers a fine contribution to a lively discussion around a topic that all humans engage willingly or unwillingly, today or tomorrow."

MATTHEW FOX
Author of *One River, Many Wells; Original Blessing;*
and *The Physics of Angels* (with Rupert Sheldrake)

"In her book, *Death Makes Life Possible*, Marilyn Schlitz, PhD, offers important insights and tools for transforming fear-based perspectives on death into awareness-based views that support our inner growth and enhance our overall experience of life."

MICHAEL A. SINGER
Author of *The Untethered Soul*

"Far from depressing or morbid, this book offers profound serenity for the dying and celebratory comfort for the living. Dr. Schlitz's multicultural work is not about dying. It is rather a path to embrace our noble end and to enjoy a life of gratitude for being here."

DR. MARIO MARTINEZ
Founder of Biocognitive Science and author
of the bestselling book *The MindBody Code: How to Change
the Beliefs that Limit Your Health, Longevity, and Success*

"This book will open our minds and hearts to the preciousness of life and help create a culture of possibility for a subject cloaked far too long in fear and avoidance. The purpose of this book is no less than to liberate you from the fear of death. *Death Makes Life Possible* will guide you on a journey to emotional healing, personal transformation, and spiritual awakening."

HIROO AND MASAMI SAIONJI
Goi Peace Foundation

"This book is an engaging affirmation of Albert Camus' edict: 'Come to terms with death. Thereafter anything is possible.'"

SHELDON SOLOMON
Professor of psychology at Skidmore College

"Marilyn Schiltz's book *Death Makes Life Possible* takes the reader on a fascinating journey along a road rarely traveled—one that explores a multiplicity of opportunities for personal and societal transformation by pondering death and understanding the implications of death's far-reaching web of fear. One of the key messages her book conveys is that death can teach us how to live a more conscious and enriched life when we can let go of that innate existential fear we all have. The sooner we start letting death teach us, the more fulfilling our lives can be. Compassionate volunteer caregivers who serve the dying learn this quickly, and their lives are forever transformed in ways they never thought possible. Compassion is the world's elixir for the fear of death."

GREG SCHNEIDER
Founder and president of Hospice Volunteer Association

"Marilyn Schlitz's *Death Makes Life Possible* is a brilliant bricolage of interviews, observations, and insights about spiritual transformation, consciousness, and our inevitable reckoning with what Henry James once called 'the distinguished thing.' From a prodigious array of philosophers, psychologists, doctors, nurses, hospice workers, professors, rabbis, imams, evangelicals, qigong masters, shamans, killjoy skeptics, and moonstruck New Age prophets, Schlitz has marshaled a treasury of knowledge about our beliefs about death and our experiences of the immaterial world. She takes readers on a search for the great grail of all spiritual traditions—the idea that consciousness exists apart from the brain and that our lives have some larger meaning beyond what we find and assign for ourselves and that the dead aren't really dead. It's a hope-haunted undertaking full of profound insights and uncanny tales of life after life, and some momentous day each of us will know for sure whether any of it is true."

CHIP BROWN
Author of *Afterwards You're a Genius:*
*Faith, Medicine, and the Metaphysics of Healing*

"*Death Makes Life Possible* is indeed revolutionary. In many ways, it demystifies and normalizes the process of dying by inviting readers to deeply explore timeless questions such as 'What is the thread of continuity in all of this *me*?' It also highlights the costs of not talking about death. The price we pay for not talking about death includes financial costs associated with dramatic and aggressive treatments that may also erode the quality of life at the end of life. There are also significant costs related to fear and a lack of healing and closure that occur when we are unable to process the experiences of living and dying personally as well as with those we love. This is a must-read for health professionals as well as the public."

MARY JO KREITZER PHD, RN, FAAN
Director, Center for Spirituality & Healing;
professor, University of Minnesota School of Nursing

"Marilyn takes us on a journey opening the possibility, with documented living examples, of coming into an expanded and heretofore unrecognized way of living. This shift aligns us with our ability to live to our inherent potential and to be aware of both the individuality and—simultaneously—the universality of each person. Paradoxically, as her title states, death is the doorway to this expansion."

SUE STEELE
Co-editor of *Exploring Issues of Care, Dying, and the End of Life*

"St. Augustine advised, 'Let death be thy teacher' because at the time he wasn't able to refer his disciples to Marilyn Schlitz's magnificent compilation of insight and wisdom on the topic. Had he, they would have not only been able to benefit from their own introspection but from the way life and death have been understood across many cultures and time periods, all framed within contemporary scientific understanding. In short, you have access, in *Death Makes Life Possible*, to a treasure that was not available back then."

DAVID FEINSTEIN, PHD
Co-author of *Rituals for Living and Dying*

"What a courageous move by Marilyn Schlitz—to bring the topic of death out into the open to be contemplated, discussed, wrestled with, and embraced as a natural part of life. Every healthcare leader, nurse, and care provider should read this book and use it as a spring-board for rethinking and refining how we in healthcare approach death, support it, educate about it, staff for it, and experience it with our patients and within ourselves."

KATHY DOUGLAS, RN, MPH-HA
Director of the film *Nurses: If Florence Could See Us Now*

"A new view of consciousness is arising within science, in which it is seen as nonlocal or infinite in space and time, therefore immortal and eternal. This 'new' perspective is of course ancient, and reminds us that the concept of the soul is not obsolete but is alive and well. This is a crucial development, because the fear of death and total annihilation has caused more suffering in human history than all the physical diseases combined. Thus Marilyn Schlitz has performed an immensely valuable service in *Death Makes Life Possible* by remind-ing us of our eternal, immortal Self. She reminds us that immortality is not something that kicks in when we die. We are *already* immortal, eternal, and soul-like. Celebrate!"

LARRY DOSSEY, MD
Author of *One Mind;* and
BARBARA MONTGOMERY DOSSEY, RN, PHD, HNC, FAAN
Author of *Florence Nightingale: Mystic, Visionary, Healer*

"Marilyn Schlitz has manifested this timely and important book which presents a profound historical, scientific, and philosophical expertise in a brilliant synthesis that is accessible to everyone. It is beautifully written as it guides the readers through a journey of opening up our worldviews and perspectives on how to turn our fear about death into a source of living richer lives full of hope and compassion."

KATIA PETERSEN, PHD
Executive Director of Education,
Worldview Explorations Project, Institute of Noetic Sciences

"This is an important book. Schlitz shows us that the limits of human growth are not fixed and opens new windows on our capacities for self-transcendence. As both an anthropologist and consciousness researcher, she leads us through a panoramic vista of worldviews on death and the afterlife. Pulling together intimate personal experiences around dying with a careful review of the scientific evidence for post-mortem survival, Schlitz invites us to examine our deepest assumptions about human existence. She treats death awareness as a transformative practice that gives life meaning, both individually and for our shared human experience."

MICHAEL MURPHY
Cofounder of the Esalen Institute

"The tears most of us will shed about death could create a wall against life—or a ladder to the heavens. Marilyn Schlitz's brilliant book beautifully presents dozens of worldviews about death, drawing from various cultures, cosmologies, scientific research, and philosophers. As I closed the last page, I felt a metamorphosis, like the 'rip in the fabric' that is death really is only a shift in awareness, an awakening to expansion. The journey she takes us on is a pilgrimage."

CYNDI DALE
Author of *The Journey After Life*

"In this groundbreaking work, anthropologist Marilyn Schlitz offers a much-needed guidebook for understanding the profound and transformative potentials of experiences such as near-death, out-of-body, and reincarnation. While the current medical model has largely ignored these widespread phenomena, arguing that consciousness is nothing more than the byproduct of our brains, the evidence for an expanded view of human potential can no longer be neglected, as is also clearly shown in the post-materialist approach in science. This important and accessible book, in which she describes how our ideas about death define the way we live our life, is the wonderful and inspiring outcome of her personal quest. It is a highly recommended read for everyone who ever wondered about the mystery of life and what may lie beyond."

PIM VAN LOMMEL, MD
Cardiologist and author of *Consciousness Beyond Life*

"You hear people—many people—telling stories, all through the book. We see through others' eyes, to see as one, and it touches us deeply, as death does. And writing it down is profound. Readers will experience a transformation of thought and feeling about death as life."

TOM JANISSE, MD, MBA
Editor-in-Chief and Publisher, *The Permanente Journal*

"There are many differences across cultures, traditions, and world-views. And there is one thing we all share in common: death. In this powerful book, Dr. Schlitz offers an insightful guide to understanding our common humanity and how our deep appreciation of death connects us all, regardless of our specific faith tradition. The subject and her treatment of it are both fascinating and timeless, and show that Marilyn Schlitz is in the vanguard of thinkers in the interdisciplinary field that spans spirituality, psychology, and culture studies."

MARIA VOLKOVA
Women's Program Director, Women's Task Force,
Council for a Parliament of the World's Religions

"Far beyond a self-help handbook, this book and film are a thoroughly researched combination of compass, map, log, and vessel for a deeper exploration into a most important journey: your life, now, as well as your death, sooner or later, on a not-too-distant shore. But what might be on that unknown shore? *Death Makes Life Possible* presents objective scientific information and subjective spiritual experiences along with new knowledge about our conscious relationships with living and dying. Using extensive interviews with experts and witnesses, Marilyn Schlitz takes us on a cross-cultural exploration into essence—up to the edge, over, beyond, and back. This book is for anyone looking for ways of living and dying with soul."

J. PAUL DE VIERVILLE
Professor of History and the Humanities at Alamo Colleges;
director, Alamo Plaza Spa at the historic Menger Hotel

DEATH MAKES **LIFE** POSSIBLE

## ALSO BY MARILYN SCHLITZ, PHD

*Death Makes Life Possible,* a film featuring Deepak Chopra, Marilyn Schlitz, and many other leading scientists, anthropologists, philosophers, and spiritual teachers. To watch a trailer or to purchase the film, visit deathmakeslifepossible.com.

*Living the Noetic Life: Transformation and Healing at the Convergence of Science and Spiritual Practice* (audio program)

*Worldview Explorations: Facilitator Guide and Workbook* (with Katia Peterson and Cassandra Vieten)

*Living Deeply: The Art and Science of Transformation in Everyday Life* (with Cassandra Vieten and Tina Amorok)

*Consciousness and Healing: Integral Approaches to Mind-Body Medicine* (with Tina Amorok and Marc S. Micozzi, MD, PhD)

*Research in Parapsychology* (with Nancy L. Zingrone, Carlos S. Alvarado, and Julie Minton)

*Reflections on Medina Lake: 1912–1987*

# DEATH MAKES LIFE POSSIBLE

## REVOLUTIONARY INSIGHTS on LIVING, DYING, and the CONTINUATION of CONSCIOUSNESS

**MARILYN SCHLITZ, PhD**
Foreword by Deepak Chopra, MD

**SOUNDS TRUE**
BOULDER, COLORADO

Sounds True, Inc.
Boulder, CO 80306

Cover and design by Jennifer Miles
Book design by Beth Skelley

Printed in the United States of America

Library of Congress Cataloging-in-Publication Data
Schlitz, Marilyn.
  Death makes life possible : revolutionary insights on living, dying, and the continuation of consciousness / Marilyn Schlitz.
     pages cm
  Includes index.
  ISBN 978-1-62203-416-1
  1. Death—Psychological aspects. 2. Future life. 3. Consciousness.
  4. Spirituality. 5. Loss (Psychology)  I. Title.
  BF789.D4.S334 2015
  155.9'37—dc23
                                    2014042617

Ebook ISBN 978-1-62203-453-6

10 9 8 7 6 5 4 3 2 1

It is truly a great cosmic paradox that one of the best teachers in all of life turns out to be death. No person or situation could ever teach you as much as death has to teach you. . . . Learn to live as though you are facing death at all times, and you'll become bolder and more open. If you live life fully, you won't have any last wishes.

MICHAEL A. SINGER, *THE UNTETHERED SOUL*

# CONTENTS

FOREWORD  Life Is a String of Beads, by Deepak Chopra, MD . . . ix

Introduction . . . xix

CHAPTER 1  Transforming Our Worldviews . . . 1

CHAPTER 2  Facing the Fear of Death . . . 21

CHAPTER 3  Glimpses beyond Death and the Physical World . . . 43

CHAPTER 4  Cosmologies of Life, Death, and Beyond . . . 67

CHAPTER 5  Science of the Afterlife . . . 97

CHAPTER 6  The Practice of Dying . . . 121

CHAPTER 7  Grief as a Doorway to Transformation . . . 131

CHAPTER 8  Dreaming and the Transformation of Death . . . 151

CHAPTER 9  Transformative Art . . . 167

CHAPTER 10  Life, Death, and the Quantum Soul . . . 183

CHAPTER 11  Healing Self and Society . . . 199

Conclusion . . . 211

Acknowledgments . . . 215

Interview Participants . . . 219

Notes . . . 223

Glossary . . . 231

Index . . . 239

About the Author . . . 245

# LIFE IS A STRING OF BEADS

## DEEPAK CHOPRA, MD

I have thought about death since the age of six, when my grandfather passed away suddenly. The event was shocking because my little brother and I had spent the day with him—a particularly wonderful day that included going to see *Ali Baba and the Forty Thieves* at the movies—and he died in that night. I woke up to hear the women of the house crying out in grief, a terrifying experience.

It's taken me almost a lifetime to fully comprehend that death is what makes life possible. The passing away of one form allows a new one to emerge. This is a continuous and totally necessary process. You were a child, and now that child is no longer here. You were a teenager, and now that teenager is no longer here.

Death offers us the opportunity to return what was given to us at the moment of birth: an invisible gift of potential. We return the gift by having actualized the potential into experience. But as an experience ends—arriving at extinction—we are always realizing more potential. We are the potential of all that was, all that is, and all that will be. That's what it means to be alive, connected to a source that keeps replenishing itself. If you can ground yourself right this moment in this eternal potential, which is your fundamental state, then life is a gift and death is a gift.

If you understand that everything in the universe, including your own body-mind, is an activity, and that birth and death are space-time events meshed together in that activity, then both are part of something greater. What is that something in which all events occur? It is consciousness. This is the answer given by the world's wisdom traditions, which at bottom are not religions but explorations of human awareness. Where is consciousness? Who is having the experience of being conscious?

These seem like enormous questions, so let me ask you a simpler one: What did you have for dinner last night?

Let's say you had pasta. And now as you remember that experience, you have a mental image of the meal, along with the people you were with and the room you were in. I can be sure that you're reliving that experience through mental images; you might even be tasting the food, seeing its colors, hearing the conversation at the table. However, if I were to go into your brain, I wouldn't see a picture that depicts what you're seeing or hear any sounds that you're hearing. The brain is totally silent and dark. I would detect only electrochemical activity—nothing like the total experience of life that you're having right now.

So where can we locate that picture, those sounds, those tastes, and where was the electrochemical activity corresponding to the picture before you decided to remember it? Any logical answer would confront a mystery. There is no location for experience in the brain before it is conjured up by an intention. But when I ask you to remember what you ate for dinner, you spontaneously have an intention, which by itself retrieves a memory. The memory was not an electrochemical memory; it was a memory of pictures and sounds and conversations and maybe the taste and smell that, taken all together, constitute your experience.

That was the memory, but it wasn't actually born until I asked you a question. So what did it exist as before that? It existed as a kind of invisible "possibility wave" in this thing we call consciousness, which has no location. Not just memories, but every experience has this double existence, first as a possibility with no location and second as an event in space and time.

With this knowledge, death looks very different from the complete cessation that everyone fears so deeply. What happens to us after we die? We go to the same place where the memory of pasta was before I asked you the question. In fact, you and I are there now—we don't have to wait until the moment of death—because existence is a field of possibilities.

An everyday example of this is very close to you: your vocabulary. You have a fund of all the words you know, which amounts to tens of thousands for an educated person. But at this moment, your vocabulary is nowhere in your brain. There is no physical trace of words stored in the brain. Even in your consciousness, your capacity to speak doesn't exist as words. Words exist in the same place as those potential memories we've been talking about. A physicist would call this place *the quantum field,* a term that has a very long tradition, one that predates the arrival of science. Lord Krishna says in the Bhagavad Gita, "I am the field, and I am the knower of that field."

The field in both cases is a consciousness field. The Sanskrit word for field is *kshetra.* And then the knower of the field is *kshetrajna.* So we are both the field and the knower of the field, because Lord Krishna speaks for all consciousness. To use another phrase of Krishna's, we curve back in ourselves to create the experiences that we call memory, which are embedded as potential. What is this curving back? Self-awareness, the most basic property of consciousness. There's a Sanskrit word for how memory is stored and activated: *sanskara.* Sanskara is not a memory; it's the seed of a memory—in other words, a potential memory.

The way in which a potential is able to emerge as a particular event corresponds very beautifully to what in physics is called *the observer effect.* Before you observe a subatomic particle, it exists as a possibility wave. Then you have the intention to observe it, and there it is, a specific isolated photon or electron. Thanks to the double life that prevails at every level, an isolated particle tells only half the story, because whether it is emerging or disappearing, a particle never leaves the field, just as a wave on the ocean, even as it rises and falls, never leaves the totality of the sea. You are in that field right now, participating in its doubleness. The next word you say, which will be an isolated event, depends on you retrieving it from the field.

Are you afraid to go there? No. We can surmise, from the way existence operates, that you will return to the same field after you die. Nothing in nature indicates that the continuum ever ends. Once you understand existence, so-called nonexistence is never the same again. This is true of the physical world as a whole. When I say "This is a table" or "This is my body," I'm experiencing colors and sounds and tastes and textures. In modern language, they're called *qualia,* for qualities of consciousness. Everything I call the objective world "out there" is constructed out of the qualities of consciousness within myself.

If I say "within myself," who is me? If you go inside my brain, you won't find me. Because the sensation "I" is also one of our qualia. It has to be, because reality is only known through experience, and experience is a mass of ever-changing qualia. As thoughts, sensations, images, and feelings flow through me, they form everything I know or can possibly know. Whatever may exist outside qualia, the human brain cannot experience it. Like the sensation of "I," the sensation "you" is also a quale. They are both in the field.

Do you remember what you were doing three weeks ago on Thursday? You most likely don't. We remember only those things that are practical or that have emotional significance. If three weeks ago on Thursday, you fell in love, you'd be a hundred times more likely to remember the date and what was happening. Emotions are very strong qualia of consciousness. Emotional thought forms are the qualia that keep the bonds of life going. It's a continuum of linked qualia pictures: the past exists as potential that I can retrieve, the future is potential that I can create, and at this moment, I'm free of both.

If you realize that you exist in this continuum that unfolds as the eternal now, then it creates no fear to realize that you're dying every moment to be reborn the next. Death is a creative process that deserves to be celebrated, not feared. That's what I really mean by "death makes life possible," because by participating in the bubbling activity of the field, where creation must embrace destruction, you're really living. Otherwise, if you're living, as most people do, either in the past or in the future, you're living in a dream.

In fact, most people spend 99.9 percent of their entire lifetime in a dream. What they're obsessed with is thoughts about the past and thoughts about the future. The one is a memory of pain and pleasure; the other is an anticipation of pain and pleasure. The Buddha warned that this obsession has no end and no escape. You will be constantly trying to avoid pain and pursue pleasure without realizing that the two are yoked together. If you want to really live, you have to be fully alive to this moment, which means being dead to the past and the future. At that point, you flow eternally in the field of potentiality.

Being immune to the allure of the past and the future is a good practical description of enlightenment. We're living in a dream unless we are awake here and now, in the moment. We assume that life is embedded in the physical, but it's not. We're perceiving only what our thought forms are allowing us to perceive.

When I die, I stop retrieving the stream of information gathered by the brain. It's an incubation time for a new emergence, a new life—however you define it. There are many possible levels of incubation, as there must be, since all worlds, not just this one, are creations of consciousness. Physicists say that on a quantum level the time between activity and nonactivity, the gap between existence and nonexistence, consists of only a few microseconds. Your body's cells die on their own pre-timed schedule depending on where they are: Red blood corpuscles take a hundred and twenty days to die, and then new ones are born. Stomach cells die every five days, skin cells every thirty days. But the body exists beyond this physical flux, maintained by memory, an invisible intelligence at the level of the field, where the real blueprint of life exists.

So the reincarnation of memories is happening right now in my body at every level—quantum mechanical, cellular, molecular. In fact, if you stop this coordination between life and death, what you get is a cancer cell. A cancer cell has forgotten how to die. It has forgotten the memory of wholeness. Having lost the memory of wholeness, it goes on its personal quest for immortality. It kills the rest of the body and then winds up dying anyway. Nature would implode into nothingness if it lost the knowledge that death makes life possible. We'd all be mummified in a frozen universe. What makes the universe fresh is that it's continually dying and being reborn. Every time it's reborn, it creates a better version of itself. We call this process evolution.

We all are a part of cosmic evolution. Our own dying, at every level, isn't static; we re-emerge, from moment to moment, as a better version of ourselves, hopefully. That's why death is so exciting—with every death there's a new opportunity. The universe can't afford to ignore each and every new possibility, which is why it recycles matter, energy, and information. Nothing is created or destroyed, only transformed. Why wouldn't consciousness be recycling itself? Why would consciousness be the exception? In reality, the field of consciousness is what recycles as energy, information, and matter.

Some years back, my father died suddenly in India, while I was far away in America. I was shocked; I felt remorse and grief over not being with him. This is the impact that death has for us personally—we grieve for what it has taken from us. But loss depends on your perspective. In that state, I asked myself: Where was my father when he was alive? He was in my consciousness. Where was my relationship with him when he was alive in the physical body? The love that I felt, the connection, the bond, were all in my consciousness. Where is he now? In my consciousness.

I believe that the key to the conquest of death is to find out who you are. As long as you identify with your body and your mind, you're being bamboozled by a superstition, because the body and the mind are dying all the time. I don't have the same thoughts that I had when I was a teenager. I don't have the same personality. I don't have the same emotions. So what is the thread of continuity in all of this "me"?

There's the continuity of personal memories. Imagine you have a strand of beads. Think of the beads as memories, which are continuously linked on an invisible string. That's you—you are the string on which memories are strung. If you can go to that deeper level, which has no memories but gives them continuity, you have conquered death. Seen as a continuum rather than as a sequence of memories, existence is impervious to death. You realize that death is an illusion; it has been an illusion all the time.

I look upon death as a quantum leap of the soul, which uses the same karmic software—memories, experiences, imagination, desire—to reinvent itself in a new context. One is reinvented in a new place, with new meanings and new relationships to continue the soul's journey. I don't base this conception on religion or faith. It describes the quantum leap that happens whenever I have a new thought or my body produces a new cell. Death is the "off" position between the "ons." The universe is a complex, multidimensional vibration, in which everything is going on and off. If the universe reinvents itself by going off, we do too.

It's important not to identify with the mechanism, however, because the self isn't a mechanical creation. You are not your brain, and you are not your body. You're the user of your brain and your body. Every time you have a mental event, there's a neural representation of it. You can see where it's happening by the spark of electrochemical activity that shows up on an fMRI. But what you are seeing is a self being actualized, the same way that the brain actualizes memories. No one has ever found the traces of memory in a brain cell. Nor are your imagination, your desires, your intentions in a brain cell. Nothing that makes us human is there. It is futile to dissect neurons to discover the location of insight, intuition, and inspiration. These are the qualities of our soul, and the soul is not locatable, even though all experiences of the soul must register in the brain, since the brain's function is to manifest possibilities of every kind, making them local, personal, and perceptible.

There are two views about consciousness in science today. One is that consciousness is an emergent property of the brain and, therefore, also an emergent property of evolution. That's the materialist, reductionist view. There's another view that's creeping up now with some postmodern scientists, inspired anew by philosophy, spiritual traditions, and evolutionary theory. This view holds that consciousness is not an emergent property but inherent in the universe. In fact, evolution is being driven by consciousness. The two views can be simplified as "matter first" and "mind first." I've been giving a "mind first" argument, where consciousness is a field effect; it's nonlocal, transcendent, eternal, the ground state from which everything emerges. Being prior to time, cosmic consciousness is not subject to birth or death.

One of the glaring problems for the "matter first" position is this: when you set out looking for consciousness, it's consciousness that's doing the looking. By definition, that's a subjective experience. You can't get around this by focusing on the brain, although neuroscience is convinced that one day a fine dissection of neurons and a total mapping of brain activity will deliver the secret of consciousness. It can't, however, because for the brain to be the source of consciousness, you must find the exact point where molecules learned to think. A sugar molecule can't think when it's sitting in a sugar bowl. At what point, as the sugar molecule is ingested and travels to the brain, does it magically learn to think?

Yet the reverse activity, thoughts turning into molecules, happens in the brain all the time. Every impulse of thought, in order to manifest, must create a unique set of chemical reactions. This is true whether you are hoping for a date tonight, feeling hungry for a baloney sandwich, or having a vision of angels. Mind activates matter by turning an invisible potential into an organized physical event in the brain. Any objective "proof" for mind derived from looking at the brain is at best inferential, the way a deaf person might look at the rise and fall of piano keys and infer the existence of music. Such proof is not direct. So you need to combine the data of science with the insights of the great sages to fully understand consciousness in its double life as both potential and manifestation. Without the potential, there is no manifestation.

Every explanation that the ancient wisdom traditions come up with in terms of where we go when we die has some validity, because the afterlife is a realm of projection. If we buy into the illusion of a physical world, if we mistake existence for getting the minimum of pain and the maximum of pleasure, then heaven upgrades the illusion. Hell downgrades the illusion. After all, the by-product of a projection must also be a projection. As long as you understand that the projection created by the five senses is not the reality, all versions of the afterlife exist on a level playing field. They are valid in the way a good movie is valid.

Our experience of God also has many versions projected from different states of mind. When we're scared for our survival, God is the one who's punishing us or protecting us. When we have peace, God is redemption. When we are creative, God is the creator. When we have archetypal consciousness, God is the worker of miracles. Every definition of God comes from a state of awareness. I think

God is actually the invisible organizing and creating principle that exists as pure potential, the source of everything, before anything manifests. But to validate this, you have to experience it. Prayer and meditation, profound love, and all the things we've spoken about can give you this experience. You set out to nurture your soul, so to speak.

Which leads me with great delight to this book by my dear friend and colleague Marilyn Schlitz. Marilyn is a gifted polymath and an exemplar of the postmodern approach to consciousness. As an anthropologist, she brings wisdom and respect for the world's cultural traditions about death and what happens after death. As a soulful scientist, she has made it her mission to engage the great mystery of transformation, of which death is an essential part. Her pioneering research on the science of consciousness provides a solid base for the mercurial study of who we really are. As a healer, she speaks directly to a fundamental source of suffering and seeks to guide us toward a new cure. And as a spiritual practitioner and gifted teacher, Marilyn leads us by example, sharing glimpses of her own transformation through the role of cultural scribe.

The vision of *Death Makes Life Possible,* along with the feature film by the same name, is to help liberate you to move beyond the fear of death. I share with this gifted author a vision for a world in which we realize our fullest potential. May you find joy, engagement, and deep contentment by reading this remarkable work. And in the process, may you awaken to your interconnections with all of life.

# INTRODUCTION

death *noun* \'deth\: the end of life: the time
when someone or something dies. Christian
Science: the lie of life in matter: that which
is unreal and untrue.

life *noun* \'līf\: the ability to grow, change,
etc., that separates plants and animals from
things like water or rocks.

What is death? What happens after we die? And how do our answers to these questions impact how we live our lives?

My encounter with these timeless questions began before I can even remember—and before I was able to formulate the queries. I was an inquisitive, precocious, eighteen-month-old toddler, wearing pink pajamas and exploring the world around me. In a careless moment, my father had left a can of lighter fluid on the yellow Formica kitchen table. I grabbed the can and put the opening in my mouth. For months after, my small body wrestled and rested in a hospital as my lungs sought the affirmation of breath, and I struggled in the gray zone between living and dying. After several rounds of intensive care, I survived. I feel sure that this experience planted within me the seeds of respect and appreciation for the art of healing. And knowing that I'd survived such a harrowing experience as a young child gave me a curiosity about the semipermeable membrane between life and death.

I grew up in Detroit in the 1960s and '70s, a time when the United States was at war with itself. It was a war of race, of class, and ultimately of consciousness and worldview. Coming of age in such a complex time and in a setting that fueled my rebellion at individual and social levels, I was alive with confusion, anger, and a desire for change.

One night, when I was fifteen, I was on the back of a motorcycle with the wrong person at the wrong time and in the wrong place. A drunk driver pulled out of the parking lot of a bar, without his car lights on, and hit the motorcycle. The impact threw my body into the air. During what I now understand was an out-of-body experience, I watched my physical being tumble through the sky and crash to the ground. I clearly recall feeling my awareness transcend my body and look down on it from a higher vantage point.

I was taken to the emergency room with a deep, wide cut in my left leg. Waiting for my parents, who were hours away, I heard talk of possible amputation. The emergency medical team did their best, putting sixty-six stitches below my knee and eventually sending me home with question marks about my leg's recovery.

Lying on the couch in my family's home during the following week, I somehow got the idea that I could and should visualize my immune system healing my leg. I lay there for long periods of time, feeling the tingles of healing. I didn't come from a medical family, and I have no memory of having heard about mind-body medicine before that time. Now I can see that I had a direct, intuitive understanding about what I needed to do to bring about my own healing.

Over the years, my worldview has expanded to accommodate a greater range of human possibility. Today, I have two well-positioned feet on the ground and an awareness that some aspects of myself are more than just physical. I have also been at the bedside of family and friends who have crossed to the other side. This has helped deepen my own understanding of mortality. I have felt the pain of loss and the transformational potential of finding peace in grief.

## DOORWAYS TO DIALOGUE

Pursuing answers to questions of consciousness and worldview has become the defining work of my life. For decades, I have researched the hidden dimensions of being human. I have conducted myriad experiments exploring the subtle reaches of mind and the existence of consciousness beyond the body.[1] I have addressed fundamental questions about the healing potentials that lie within our ability to change and transform. I have questioned many thousands of people, including average folks, children, accomplished scientists, and acclaimed wisdom holders who represent many traditions and worldviews.

I have worked with teams of colleagues, including psychologists Cassandra Vieten and Tina Amorok, to develop a model explaining how our worldviews transform. That worldview transformation model draws on the nature of experiences that cross the distance between our physical and our metaphysical knowing and being. We published our preliminary findings in the 2008 book *Living Deeply: The Art and Science of Transformation in Everyday Life.*[2] Left out of this earlier work was the role that death awareness plays in how we live and how we transform. In my own process of exploration, I have sought to understand the role of death in our personal growth, healing, and spiritual awakening.

## THE COSTS OF NOT TALKING ABOUT DEATH

The topic of death is both timely and relevant to all of us. As the demographic facts reveal, we're an aging population. There is an unprecedented increase in the average age of people throughout the world. In 2008, the number of people worldwide who were sixty-five and older was estimated at 506 million; it is calculated to hit 1.3 billion by the year 2040. The population of the United States is predicted to reach 400 million by 2050; approximately 20 percent of people will be sixty-five years old or older. The baby boomers, of which I am one, are finding themselves at retirement age, despite their best efforts to stay young; more than ten thousand baby boomers reach the age of sixty-five every single day in the United States. This group represents the fastest growing segment of the US population.[3]

Cornell University social psychologist and boomer Daryl J. Bem told me, "Given my own age—I'm now in my seventies—I have shifted, the way that many older people do, from the thought of one's life up until this point, and instead, [to] the thought of one's life until death. And that shifts one's perspective in many, many ways." Like Bem, boomers are starting to think about their own death and what may lie beyond this life.

America's boomers have long been characterized by their individualism. Today this vigorous autonomy is challenged as we are confronted with the care of aging parents and ill or disabled children or spouses, as well as with our own mortality. We are looking for innovative ways to redefine our changing identities, roles, and responsibilities. As we age and confront our own existential issues, many of us are seeking new sources of meaning and purpose. Many of us are pursuing a self-reflective quest for wholeness, exploring diverse practices and approaches that allow us to forge our own truth system, alone and in the company of others. Some are returning to their faiths of origin. Others are embarking on a new spiritual path that may help them to live more authentic lives. For boomers and beyond, there is an expectation that old age can be better, and so we are open to developing new skills and new ways of aging gracefully.[4]

Despite how relevant the topic of death is for everyone, many of us do not like to think or talk about it. If we are healthy, we may be less likely to prepare for its inevitability. If a member of our family is terminally ill, we may be reluctant to bring up the subject because it means acknowledging the truth of the situation or offending our loved ones. Many of us hesitate to talk about death due to our own fear or our culture's taboo about discussing death. But by not talking about this potentially charged topic, people miss the opportunity to share with their family and friends about their wishes and hopes. We give up our autonomy, decision-making power, and personal authority. In this process, we all deserve better.

Unfortunately, our reluctance to contemplate death is creating significant problems. According to a recent survey by the California Healthcare Foundation, six out of ten people say they don't want their family burdened by end-of-life decisions.[5] At the same time, nearly 56 percent of the 2009 survey respondents have never communicated their end-of-life preferences to family members. For example, most Americans want to die at home, yet only 24 percent of people over sixty-five are able to fulfill this wish. Many people find themselves in nursing homes or hospitals in their final days. End-of-life care in hospitals may translate to high-cost, dramatic, and aggressive treatments that erode the quality of life. A 2010 Dartmouth study found that more than 40 percent of elderly cancer patients were treated by ten or more physicians during their last six months alive.[6] This same report found that many of these patients had some intrusive, life-prolonging procedure in the last month of their life. According to a 2010 article in the *Journal of Palliative Medicine,* only 15 to 22 percent of seriously ill elderly patients had their wishes recorded in their medical records.[7] The Agency for Healthcare Research and Quality reports that as many as 65 to 76 percent of physicians were unaware that their patients had recorded their end-of-life plans.[8]

In America, one out of every four Medicare dollars is allocated to those who are in their last year of life. Out-of-pocket expenses exceed the families' financial assets in 40 percent of households, according to a Mount Sinai School of Medicine survey.[9] Valiant efforts are made to delay what is often inevitable. For many families confronted by a medical crisis, heroic measures are a price worth paying in order to reduce the suffering of their loved ones. Unfortunately, their loved ones may not suffer less. In some cases, treatments may lead to greater harm and suffering, and the best possible care may be no treatment at all. In the light of these challenges, more and more people are seeking to help their dying loved ones end their lives through dignified, peaceful, and compassionate means.

## HEALING OUR COLLECTIVE WOUNDS

We need to address and heal the great denial we hold over what is an inevitable truth. This means diving into our worldviews about mortality. Any discussion about how we'd like the end of our lives to unfold raises questions about what we think may be coming after death. What death means for living and what happens after are big questions that unite everyone. People from all ages and walks of life are seeking answers to these same questions.

Different cultures, worldviews, and belief systems offer different perspectives on mortality and the nature of human existence. In my own search for answers, I have found myself at many doors. These included the majestic carved doors of Grace Cathedral in San Francisco, the simple wooden doors of a Sufi mosque in inner-city Oakland, the gilded doors of a Buddhist monastery in Taiwan, the screened door of the primate habitat at the Oakland Zoo, the automated doors of a high-tech surgery room in Tucson, and the daunting steel door to a two-thousand-pound, electromagnetically shielded brain-monitoring laboratory in Petaluma, California. I conducted hundreds of hours of interviews. I convened focus groups, participated in ceremony and ritual, and collected data and stories. Each step of the way, I sought to identify core threads in the complex, multidimensional tapestry of life, death, and what may lie beyond.

Poignant insights have arisen for me as I've learned how people from a variety of backgrounds embrace the challenges of their own impending deaths and the deaths of their loved ones. I have been inspired by how they have grown from such experiences. Moving into the mysteries of our mortality, I have come to see the ways in which death connects us to all of life. Through intimate and tender conversations, I have been moved from tears to awe to deep belly laughs, all in a matter of moments. I have witnessed my own personal transformation and healing in the process.

My goal in this labor of love has been to see beyond any one approach to death. Instead, I've focused on common elements that emerge when we look across a rich variety of worldviews, belief systems, and cultural perspectives. These elements are the glimmerings of a cosmological pattern. At the same time, my goal has been to affirm, support, and respect the diversity of religious, spiritual, scientific, academic, and social points of view. My focus has been on the natural and inherently cyclical process of life and death. I've sought to make the myriad teachings accessible, without diminishing their complexity.

In the pages to come, we will take a penetrating look into diverse views of death and the afterlife and explore how these views can inform and heal us, individually and as a global society. We will meet people who are of different ages and come from diverse walks of life, religious backgrounds, and cultural traditions. We will consider the great tableau of death and the terror that people experience when they deny the questions surrounding it. We will hear about how people's direct personal experiences, which suggest broader realms of existence beyond the physical world, have helped them overcome this fear. We will learn about ways in which materialist scientists are approaching questions of consciousness beyond death and why the topic matters for our understanding of reality beyond our individual embodiment. We will ponder patterns that help illuminate new answers to age-old questions—and raise new questions from age-old answers. We will explore the realizations that are embedded in evocative questions about consciousness, death, and beyond.

Along the way, each of us has the opportunity to map our own place on the transformative path. Such a journey takes us from our own physical being to our connection to a broader reality and an interconnected whole. On our own, each of us will be able to examine our worldviews and become the cartographer of our own lived experience, as informed by a panoply of beliefs, myths, stories, and scientific discoveries. As we delve into a wide array of alternative worldviews, we have the opportunity to broaden and deepen our own perspectives about consciousness and our unique human existence. We can reflect on what gives our lives meaning and purpose, what moves us beyond extrinsically oriented goals involving material gain toward intrinsically oriented goals that connect us to something larger and give us a more expansive view of life. We may expand our worldviews by learning from our own experiences and those of others along the way.

As we consider how death makes life possible, we may become happier, healthier, and better citizens. Likewise, we may begin to renew our sense of purpose in the context of our rapidly changing times. We will lay the groundwork for a new way of understanding ourselves in the light of our own mortality and that of our loved ones.

I invite you to join me in a transformative process that may help us all to find the healing potentials that lie in our relationship to death. Let us come together in a continual learning laboratory as well as in what Jerry Jampolsky, founder of the Center for Attitudinal Healing, described during an interview as "an unlearning laboratory, to unlearn some of the things we're attached to."

I will share with you what I have gleaned, from many experts and wisdom holders, about the transformative potential of death and how our fears of death can evolve into an inspirational life value. My hope is that this book, and its companion documentary film by the same title, will help you to embrace the challenging experiences of your life, including the deaths of loved ones, your own mortality, and the endings (and, in turn, the beginnings) that emerge within your own steps along the magnificent and mysterious journey of life. In this dynamic process, we may find that our own individual transformations may help catalyze changes in our society that are more joyful, just, compassionate, and sustainable.

# Chapter 1

# TRANSFORMING OUR WORLDVIEWS

Someday you will face your own mortality. At that moment, I hope you see that your life has been well led, that you hold no regrets, and that you loved well. On that day, I hope that for you, it has become a good day to die.

LEE LIPSENTHAL, *ENJOY EVERY SANDWICH*

Lee Lipsenthal was fifty-three years old when his doctor told him he was dying. He'd been living with esophageal cancer for about two years. After a dance with remission, the disease had returned full force. As a physician who was married to a physician, Lee knew that allopathic medicine had run its course. Dying was now what he was doing with his life. While his days were ebbing, he was living each moment as if it were his last.

For a decade, Lee had served as the research director for Dean Ornish's Preventive Medicine Research Institute and had also served as president of the American Board of Integrative Holistic Medicine. Despite his scientific training, he held a strong metaphysical view that guided him toward his death. Through a deep meditation practice and shamanic journey work that expanded his sense of self, he understood that his worldview structured what he believed was coming next.

Lee and I were friends and colleagues for more than a decade. During our heartfelt talks over the years, he reported remembering significant, spontaneous past-life experiences that connected him to

God, Jesus, and Buddha. When asked how these past-life experiences with religious figures informed his experience with cancer and dying, he explained:

> There's a big story there. . . . At one level, it gives me a sense of peace that this may not be all there is. But on the other level, it's just a small part of why I'm feeling well in the process of supposedly dying. The other pieces to me are a deep appreciation for the life I've had. I've had a blast. I'm a music fanatic, and I've hung out with and played guitar with some of my favorite rock 'n' roll heroes. I've had a really good ride in my fun life. . . . I've had work that has been fun, challenging, and creative. I have thirty years of a marriage to someone I'm still deeply in love with, two kids who are really wonderful people. So if I were to die now, that's fine. I don't really need more of that. That's one reason I'm at peace.
>
> The other reason is I truly know that I have no control of whether I live or die . . . so the combination of accepting a lack of control over my own death and a very deep gratitude for the life I've already had is my real reason for being at peace. I look at people who have survived a major health crisis, and they've transformed dramatically through that crisis. Is that bigger than losing your body? I don't know.

Lee's worldview allowed him to feel a fluid connection between living and dying. It was a belief structure that gave him a sense of hope and possibility. Given his own experiences with past lives, I asked him what he thought might be coming next, after the death of his body. "We are all limited by our own experiences and how we interpret them," he explained to me. Such interpretations or worldviews may be shaped by our parents, our education, our religion, what we read. And, for Lee, they were also informed by experiences of past lives and mystical states:

I think we come back into life for new and different experiences. The purpose or the meaning of that . . . I'm not going to pretend that I honestly know. I think that we progress over time in our multiple lives, we change . . . we learn from these past lives, and we become . . . let's say, better, deeper, shinier. That's my belief structure. I'll try to let you know when I get to the other side—that's all I can say.

## UNDERSTANDING OUR WORLDVIEWS

As Lee demonstrates, our views on life, death, and the afterlife are informed by diverse and sometimes competing worldviews. Religious beliefs often shape our views of death and what happens after. And there is an evolving spirituality that combines traditional religious elements, emerging insights from science, and personal practices to help address core existential questions that influence people's beliefs about life after death. When Americans are asked the standard question posed by the Gallop Institute, "Do you believe there is a life after death?" about 75 percent say yes.[1]

As a nation, Americans have their own unique perspectives compared to other nations. In 2013, George Bishop, a professor of political science at the University of Cincinnati, reported on two cross-nation surveys for the International Social Survey Program (ISSP): "On this score Americans were more certain of a hereafter than anyone else (55%): twice as sure as people from the Netherlands or Great Britain, five times as confident as the Hungarians, and nine times as convinced of it as the East Germans."[2] My own observations reveal that, independent of religious beliefs, people throughout the world are asking deep questions about life and death. Indeed, there is a growing hunger among people from all walks of life to talk about death and what may happen after.

Our views of death may also be informed by our occupations. For example, a nurse's or physician's approach to death may focus on fighting it, keeping it at bay, and trying to use whatever heroic measures are available to sustain the survival of the body. In the armed forces, people are trained to be fearless in the face of death. Still, their own religious or spiritual orientation can inform how they see the possibilities for what may come after, providing them with a sense of hope and possibility, even in battle.

Working with animals has given Margaret Rousser, zoological manager at the Oakland Zoo, a deep appreciation for the cycle of life and the transformative nature of death. She explained:

> The circle of life may be a cliché, but it really is true. You need all aspects of life and death to have this beautiful planet that we have. When an animal dies, the carcass becomes food for vultures and for beetles. When a plant dies, the fallen leaves make the soil richer for the next plants. We really do need all of these things to live on this beautiful planet.
>
> I would say that the unnatural cycle of death would be the cycle that's sped up by things that we as humans are doing incorrectly or that are harmful. There's a certain amount of a natural progression: you grow old, you age, you pass away, and you become part of the earth again. But there are things that we're doing that are causing those things to happen too quickly. We're putting toxic chemicals into our planet that are causing animals to change and mutate. So those are things that are disrupting the natural cycles. And those are the types of things that we need to take a look at as a human race and civilization. And make those changes.

As Rousser notes, death is natural. Attempts to disrupt the natural cycle are what create pathology, within our individual lives and in our relationship to nature.

A similar view was expressed by Rick Hanson. I first met with Hanson at a small café in Mill Valley, California. The founder of the Wellspring Institute for Neuroscience and Contemplative Wisdom, Hanson is a psychologist who studies evolutionary neuropsychology.

Given his background, it isn't surprising that Hanson's understanding of death is informed by a big picture of the natural cycle of life:

> Death clears the way for young to come forward, and it enables a species itself to adapt and improve itself over time. In other words, if members of the species do not die, the species could not evolve. So we're sitting here today, 3.5 billion years after life emerged on the planet, at the top of the food chain, in some sense aided and handed off to by all of the creatures that have died before us, so that we may live here today. It's in that context that I find myself experiencing gratitude and appreciating those who have died and the role of death in life.

Hanson is a Buddhist practitioner. Although he's a scientist by training, his personal philosophy about life and death have also developed based on his spiritual practice:

> The nature of dying is a teaching about the nature of living. In other words, as we die, the body decomposes and its elements spread out, one way or the other. As the mind dies, its elements spread out, decompose, and disperse. All eddies disperse eventually. That reality of death and of dying is also the reality of living. In any moment, the body and mind are constantly changing, made up of so many parts, so many patterns emerging, coalescing, organizing, stabilizing, and then moving on. Just like the little kid says at the end [of a meal], "All gone." That's the nature of every moment. The ending of a lifespan helps us appreciate and turn more into the peace and the wisdom of seeing the ways in which every moment itself has a living and a dying.

Hanson's perspective illustrates the multidimensional nature of our personal worldviews. He holds scientific, clinical, and spiritual perspectives simultaneously. His particular worldview, in turn, inspires his approach to life:

> Starting back in my twenties, I began appreciating the wisdom of Don Juan in Carlos Castaneda's books. The teacher, Don Juan, essentially says, live in such a way that you're prepared for death at every moment. I think there's a lot of wisdom in living that way because you just literally never know. We all know people who suddenly stroked out or suddenly got cancer, and they were dead ten days later, or ten months later, or maybe ten years later. But it also is in that context that it makes me appreciate hugging my kids every night, enjoying every sandwich, relishing every sunrise. . . .
>
> If you know the movie is going to come to an end, at least as far as you're concerned, it really motivates you to make it as good a movie as possible, enjoy it as much as we can, and not ruin it for others.

## APPRECIATING WORLDVIEWS

As an anthropologist, I have studied worldviews and how they shift or stay the same.[3] Worldviews are fundamental to our lives. They are the lens of perception through which we experience, understand, and interpret ourselves and the world. They inform and are informed by our beliefs about and perceptions of our human experience and the cultural and physical environments around us.

Our worldviews influence how we adapt to changes in our life circumstances. At a fundamental level, Lee Lipsenthal's worldview provided a frame of reference for his understanding of his imminent death. It provided a context that helped shape his intentions, his actions, and his emotions. It offered him solace when he confronted his own mortality.

Developing our appreciation for the power of worldviews may be understood as developing a type of literacy.[4] Each of us can learn to better understand and appreciate our own worldview and those of others. Harnessing our capacities to listen and examine multiple perspectives with humility and curiosity can open us to new ways of being. As we explore worldviews, we have a chance to not only reflect upon, comprehend, and communicate our own worldview, but also to recognize that our beliefs come from our particular life experiences and personal values. We may learn to better appreciate that other people hold different and potentially equally valid models of reality out of which their assumptions and actions arise. This capacity to appreciate other worldviews involves the growth of our own cognitive and cultural flexibility. It compels us to draw on our deepest creativity and resilience in the face of differing points of view. Improving our worldview literacy prepares us to adapt to new insights that come when we encounter different perspectives, customs, practices, and belief systems. At the most fundamental level, worldview literacy allows us to bring awareness to what is true for each of us about life, death, and what may lie beyond.

I am captivated by the diversity of worldviews about living and dying. And I'm not alone. There is increasing interest and need for people throughout the world to understand the changing cultural and religious landscape. Record numbers of people are looking for ways to make sense of the many different and competing truth claims that coexist. It can be challenging to realize that many people are reading from different scripts, reciting from different prayer books, living in different models of reality. Yet increasing numbers of us are seeking to discover our own authentic truth and expand our consciousness by reveling in the life paths of different cultures, worldviews, and belief systems. It is the premise of this book and its companion documentary film that understanding diverse truth systems about death and the afterlife, while at times challenging, can renew our sense of connection to the whole of life. Ultimately, encountering and appreciating alternative worldviews enriches each of us.

## HOW WE VIEW DEATH MIRRORS HOW WE EMBRACE CHANGE

Consciously transforming our worldviews involves appreciating the ways in which we experience the world. Our worldviews about death are a key to understanding our own identity and what happens to it when we are no longer embodied. As we begin to understand and dive more deeply into death as a rich and complex component of life, we have an opportunity to look at our own assumptions, beliefs, and expectations. Indeed, looking at how we view death is a way of seeing how we may overcome our fear of it to live more deeply. In turn, how we view death often mirrors how we transform and change throughout our lives. Mingtong Gu, a qigong master and teacher, shared his insights on why is it important to contemplate death.

> My first response is, we're all going to face death sooner or later, so we might prepare in some way, early enough.
> Secondly, life is about . . . embracing all expressions of change—change from childhood all the way to adulthood and aging. And in the change, we're constantly learning to listen for how to embrace change. The change often challenges us to find a deeper capacity to embrace the change and the challenges that come with that. It's a deep listening we're learning here. It's part of life.
> Contemplating death can really bring us peace, knowing the impermanency that is underlying all life. Without impermanency, there is no change, there is no life. Can you imagine your life without any change? It doesn't exist in the universe, as far as I know. So the deeper level of contemplating death is about embracing change, embracing impermanency. In that deeper level we can increase our capacity to embrace life, but also find a deeper peace in the midst of all these dramas, all this impermanency, all this disappointment, and even all this excitement.

Like Mingtong Gu, Brother David Steindl-Rast teaches that we are given the opportunity to grow and change in every moment. Embracing transformation is a way of embracing life and its cyclical nature.

Brother David is both a scholar and a Benedictine monk. Born in 1926 in Vienna, he studied art, anthropology, and psychology, receiving a doctorate from the University of Vienna. In 1952, he followed his family who had emigrated to the United States for new opportunities. In 1953, he joined the newly founded Benedictine community in Elmira, New York, Mount Saviour Monastery, of which he is now a senior member.

Spiritual transformation, he told me, is what the universe wants from humans. Our incarnation allows humans to

> transcend the natural creation and dissolution of form
> and to find that point within ourselves, which is the
> watcher or the still point of the dance, if you want,
> and in that sense step out of that natural process in
> which forms come to be and then decay again. In the
> midst of all this coming and going of forms, there is
> this hunger in every human heart for that which lasts.
> If it is misunderstood, then we cling to this form or
> that form.

To make his point, he described an outing in San Francisco when he came upon the store Forever 21. The lesson for him was that if people cling to being twenty-one, they won't be happy. By resisting aging, people deny a fundamental aspect of their human experience. "You can't possibly achieve it, so you might as well go with the flow and not insist on being forever twenty-one," he said. The nature of our worldview must be fluid, not fixed at a particular place or time in our mind. He continued:

> The misunderstanding would be that we cling to one
> form in this process. The right understanding would
> be that we go with this flow because it's a given—
> we can't help it anyway. We better go with the grain,
> rather than being dragged by our ears, and at the same
> time transcend this flow of form. Go to that realm
> beyond space and time. Language cannot express it

very well. In the first stage, we observe this flow. So we are no longer identified with it. We are in the present moment and no longer caught up in the past and future; we observe it. Then the next stage of this inner development, this unfolding, would be that we allow the power and the energy that brings forth all these changes in the creation of forms and the destruction of forms, to flow through us. Realize that we are one with it, that we are really not only created, but [also] one with the creating force.

It is very clear our physical bodies are decaying and will eventually disappear, and other bodies will come. We may have something to do with it because all the raw material will be recycled and used somewhere. We are not detached, not completely separating ourselves. We are only detaching ourselves sufficiently so that we can watch it and realize that we are part of that cosmic force that drives it.

## UNDERSTANDING WORLDVIEW TRANSFORMATION

Worldview transformation involves a fundamental shift in the way we see and interpret the world. It also involves a change in how we see ourselves and our relationships to others. It's not simply a change in our perspectives, but involves a different way of understanding what is possible. It defines what gives us meaning and purpose in our lives. As we transform our views of death, we can begin to understand our own mortality in a new light. Frances Vaughan, a transpersonal psychologist who has worked at the interface of living and dying for many years, explained worldview transformation this way:

It's the capacity to expand your worldview so that you can appreciate different perspectives so that you can hold multiple perspectives simultaneously. You're not just moving around from one point of view to another, you're really expanding your awareness to encompass more possibilities.[5]

Over the past twenty years, my colleagues and I have done research on transformation. Our team of psychologists and anthropologists has collected narrative descriptions of transformations reported by people from many walks of life. We interviewed teachers and leaders who had come together in focus groups. From these conversations, we formally interviewed sixty masters, using a set of twenty pre-specified questions. We analyzed the content of these interviews to gain insights into both the similarities and differences across diverse religious, spiritual, and transformative traditions.

We found that people often described their transformational experiences as a kind of hero's journey, and that, from unremarkable to extraordinary, these experiences of epiphany and insight led to fundamental shifts in people's sense of personal identity, their relationships to others, and their embeddedness in the world.

Based on this large-scale qualitative study, we created several surveys and administered them online to over two thousand self-selected individuals. Our goal was to further explore the process of transformation and to understand how life experiences can catalyze worldview shifts that are beneficial to both individuals and society. The experiences that triggered consciousness transformations were diverse, ranging from someone suddenly seeing the world in a new way while washing dishes to someone redefining their values and priorities in the face of a life-threatening illness. And yet, as we detailed in *Living Deeply,* there were common threads that hinted at a brilliantly woven tapestry of human experience that overarched individual and cultural differences.

## The Worldview Transformation Model

Out of this work conducted at the Institute of Noetic Sciences, my colleagues and I created a model that shows the dynamic nature of worldview transformation. Figure 1 depicts this model in graphic form. I have used this worldview transformation model to help guide the organization of this book. By following this model as you read, you can engage in a path of discovery and practice that may lead you beyond the fear of death and into a deep and lasting appreciation for the precious life each of us has been given.

Our decades of research on transformation show that shifts in worldview are most commonly triggered by experiences of pain and suffering. Life events—including illness, divorce, job loss, and the death of a loved one—can disrupt the steady state of a person's life, giving them an opportunity, if they can see it as such, to alter their path and to live into an expanded, meaning-filled worldview. Painful and frightening experiences have the capacity to loosen our sense of control over our life circumstances. This can be a good thing. They can help to dissolve our fixed identities in ways that broaden our understanding of who we are and what we are capable of becoming. As physician and teacher Rachel Remen observed from her years working in the field of oncology and humanistic healthcare:

> Crisis, suffering, loss, an unexpected encounter with
> the unknown—all of this has the potential to initiate
> a shift in perspective. A way of seeing the familiar
> with new eyes, a way of seeing the self in a completely
> new way. As I wrote in my book *My Grandfather's
> Blessings,* this kind of experience shifts a person's values,
> shuffles them like a deck of cards.[6] A value that's been
> on the bottom of the deck for many years may now
> turn out to be the top card. There's a moment when
> the individual steps away from the former life and the
> former identity and is completely out of control and
> completely surrenders—and then is reborn with a
> larger, expanded identity.

Not all catalysts of transformation are associated with suffering, however. People report spontaneous sensations of deep awe, wonder, and a profound connection to something greater than themselves. These personal experiences can often encompass what have been described as mystical encounters with unseen worlds. They can also include insights, inspiration, and moments of ecstasy. These experiences

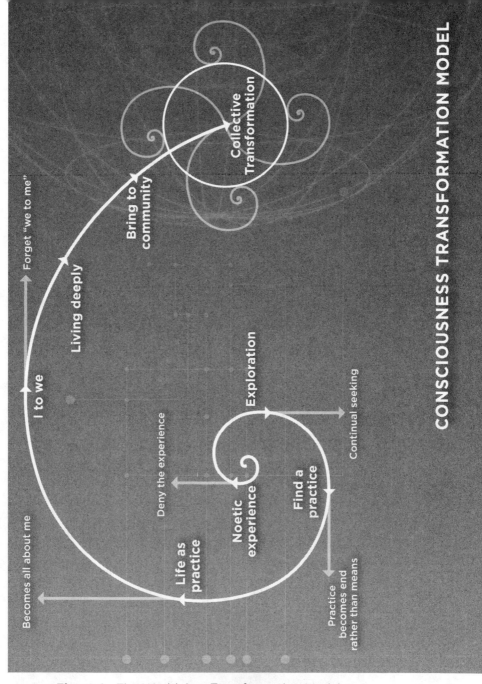

**Figure 1** The Worldview Transformation Model, created by Cassandra Vieten, Tina Amorok, and Marilyn Schlitz. (Adapted from *Living Deeply: The Art and Science of Transformation in Everyday Life.*) Available as a PDF via the Institute of Noetic Sciences website: noetic.org.

move us beyond our narrow definition of the self and into a deeply rooted, embodied sense of unity; an awareness of great love; and a fundamental sense of interconnection.

Such positive transformations may need many small glimmerings of insight in order to take hold. One of the blocks to transformation that we identified in the worldview transformation model is the tendency to deny or resist change. In such cases, numerous experiences may be needed to shift a person's worldview.

It is also possible for positive worldview transformations to occur in an instant. Sudden personal metamorphoses, called *quantum changes* by psychologists William Miller and Janet C'de Baca, may be fueled by experiences that take people out of their ordinary, steady state.[7] These experiences may include epiphanies, "big dreams," and revelations. They may also encompass experiences that suggest an extended reach of our human consciousness, including near-death experiences, spontaneous healing, or other phenomena that arise in nonordinary states of consciousness (such as seeing apparitions, sensing oneness with the whole of existence, or having precognitive dreams). Transpersonal scholar and archivist Rhea White found that even though the phenomenology of such experiences may differ, all can serve as portals to a new worldview that embraces oneness with all of life, interconnectedness, and prosocial values.[8] In chapter 3, we will consider the implications of such experiences for understanding death as a catalyst for worldview transformation and healing.

These breakthrough experiences can be puzzling for people. While there may be no external "proof" to validate or explain many such experiences, they can be powerful and life-transforming. To make sense of these transformative triggers, we enter a stage of exploration and discovery. Learning that we are not alone in our intention to transform can be very reassuring. In this book, we will consider diverse paths of exploration that we can use to make sense of intuitive insights and revelations and to heal our fragmented lives. These include insights from diverse cultural, religious, and spiritual traditions; science; and a variety of transformative practices that lead to direct personal experiences with nonordinary consciousness. We will also consider evidence for an emerging new worldview that links inner wisdom with the science of entanglement and quantum physics, leading to the emergence of an evidence-based spirituality.

## Transformative Practices

In the process of exploration, people often discover practices that can support them in their own transformative process. This is the third phase in the transformation model. These practices may serve to ground and sustain transformative insights so that they may be integrated into our everyday lives. Through formal and informal practices, we can begin to internalize new understandings. In this way, we redefine our sense of identity and how we fit within the broader world in which we live. Engaging in transformative practices may create space for other aspects of our lives—including satisfaction, happiness, creativity, intuition, compassion, and altruism—to take their places in how we express our authentic selves. The practices may help remove blocks that keep us encumbered by the fear of death, allowing us to embrace a new way of holding our life.

There are myriad forms of transformative practice. Some are ancient and based on well-established rituals and stages of development; others have emerged more recently and are less structured. But all may help to transform the fear of death into a life-affirming value. This book considers how specific tools—including death preparation, guided imagery, meditation, prayer, grief, dreams, and art—can help us transform our understanding of death and what it means to be alive.

The worldview transformation model identifies five elements that are common across many transformative practices.

> **Intention:** setting an intention to transform the fear of death into an inspiration for living.

> **Attention:** shifting where you place your attention to include dimensions of life that have previously been ignored and that reveal new paths of understanding about life, death, and what happens after, offering you new meaning and purpose.

**Repetition:** building new habits or ways of thinking and behaving around your understanding of mortality, through repetitive actions that can open you to new perspectives, values, and actions supporting a well-lived life.

**Guidance:** finding guidance about your worldview on death and your intentions for growth and transformation from diverse sources, such as trusted authorities, community programs, personal relationships, and quiet reflection.

**Acceptance:** surrendering your sense of control over your life circumstances to help dissolve any fixed identities you have that limit you, and to expand your understanding of who you are in the broadest sense.

As we seek to transform our views on death, we can build new perspectives on our mortality that redefine who we are and what we are capable of becoming.

There are limitations to transformative practices. When we view our practice as an end in and of itself, we may begin to see ourselves as special or virtuous. We run the risk of viewing our commitment to a transformative path as something that places us above others. In this process, we can become trapped by our own egos and lose sight of our connection to something that transcends our individual identity. In such a case, we may be seduced into thinking that the experience is all about "me." It may also be that we identify practice with a particular context or form, failing to see how transformation applies to our everyday experiences. As my colleagues and I point out in *Living Deeply*,

> at some point in the transformative process, you recognize that there's no difference between who you are in the pew or on the aikido mat, and who you are in the grocery store, on the freeway, or at your office. The same mindful attention brought to the placement

of your legs in a difficult yoga pose can be brought
to a challenging conversation with your child. The
same peace and joy brought to a beloved community
of fellow practitioners can be brought to a PTA
meeting. The same reverence that arises from spending
three days in the wilderness on a vision quest can be
brought to the clouds in the sky and the spindly trees
in the mall parking lot.[9]

If we can avoid the pitfall of regarding transformative practice as an
isolated activity done on certain days and at certain times, we may
begin to find that life becomes practice. In this process of engaging in
a transformative practice, we may observe a shift from "me" to "we,"
from focusing primarily on our own individual needs and desires to
transcending through them to a more inclusive worldview.[10] At the
same time, it is not selflessness, but a more authentic sense of self that
emerges. Recognizing our interconnectedness can give us a larger
framework in which to see ourselves in relationship to others—both
living and departed—and can lead us to the spirit of service to
others. Transforming our fear of death can result in the development
of greater kindness, generosity, love, compassion, forgiveness, and
altruism. Further, it can expand our sense of community and make
it more inclusive, as we drop the illusion of the "other" and more
fully recognize our fundamental interdependence with all of life. We
may notice that we become less reactive to people whose worldview
is different from our own, regarding them instead with curiosity
and appreciation for our common path as mortal beings. As we will
explore in the following chapters, overcoming the fear of death can
help us to live more fully and deeply in our rich multicultural world.

Finally, we consider how worldview transformation can be applied
to the development of more sustainable social practices. Revising
the assumptions and the policies that guide our shared social and
cultural values around death may lead us to full-system change. As a
critical mass of people is able to transform their collective pathology
around death, we may revise how our social institutions, in particular healthcare, treat death and dying. We may achieve a tipping point

that serves our collective wellbeing. In this way, we may integrate our own inner transformation with the evolving cultural landscape of which we are co-creators.

## GLEANINGS

Experiences that challenge our assumptions about death and an afterlife may open us to new worldviews or ways of experiencing our lives. Paying attention to our own direct personal experiences, including deep intuition, epiphanies, and expanded states of consciousness, may help us to reduce the fear and suffering that many of us experience about death. Intentionally engaging in the transformative process may lead to both subtle and dramatic shifts in our understanding of who we really are. We may begin to fully understand ourselves as healthy individuals and as parts of an interconnected whole that knows no boundaries. Through various experiences and practices, we may witness a transformation that moves each of us from isolation to belonging. Understanding our relationships with the seen and unseen realms of everyday existence offers us new possibilities for living a big, juicy life filled with humility, purpose, meaning, and hope.

‹ PRACTICE ›

### Exploring Your Worldview in the World around You

In this chapter, we have considered the role worldviews play in shaping our views of reality and how, over time, these views can transform in ways that enhance our lives. The following practice offers an opportunity to observe your own worldview and how it shapes (and is shaped by) what you see about the world around you.[11]

To begin, spend five to ten minutes in a place that you visit often. It can be a store, a park, your neighborhood, a local bookstore, a restaurant, your own backyard. Take a slow walk through the place you've selected. Observe your surroundings as if you've never been there before. Be curious about what is happening in this place, paying special attention to things you have never observed before. Bring your awareness to your senses, noticing what you hear, see, smell, taste, or feel.

Next, find a quiet place to sit and journal. Take a few deep breaths, clear your mind, and center your body. Reflect on what you just experienced. Were there things about the place that were surprising, even though you'd been there before? Did anything make you curious to learn more? Did you notice that you had certain expectations about the place and what would happen there? Did you find things that did not fit your expectations? Did things change when you slowed down, brought your intention to the practice, and paid attention in a new way?

Consider the ways in which your experiences in this familiar place are informed by your worldview or the lens of perception through which you experience the world. Write in your journal for at least ten minutes, recording what you observed by visiting the familiar with fresh eyes. Learning to notice your own worldview may open you to new ways of experiencing your life and your relationship to death.

# Chapter 2

# FACING THE FEAR OF DEATH

If you want to love life, that means loving death.

TONY REDHOUSE

Why, we might ask, is there such a fear around death? What has made death the great taboo topic, in spite of the fact that we all die? Josh, a bright thirteen-year-old boy, has given death some thought and offered this insight:

> I know I'm afraid of death because I don't want
> to think a different way. I don't want to become
> a different person. I just want to stay who I am.
> If I change, I want to remember this form or, I
> guess, person.

The famed writer Mark Twain suggested another reason: "The fear of death follows from the fear of life. A man who lives fully is prepared to die at any time."

Luisah Teish, an Oshun chief in the Yoruba Lucumi tradition, echoed both Josh and Twain. She observed that modern society has, in many ways, estranged us from our natural role in the cycle of life.

> Everything is constantly being transformed. I don't
> see any evidence that death is just the end. But I
> think that whoever controls resources, media, images,

and education can cause people to come to fear and hate the natural cycle. The fear of death is an attitude that the media has sold us. It bounces between fear and romanticism.

I personally have more fear of an unfulfilled life than of death itself.

As Teish noted, our worldviews about death are informed by many factors. While such factors can bring death to our attention, it is equally true that many in the modern industrialized world have rarely seen a dead body. As Daryl J. Bem, a social psychologist from Cornell University whom we briefly met in the introduction, pointed out:

If you ask people in the advanced industrial world like the United States, very few people have actually seen a dead body until it's been prepared, and even then you often don't see them until your grandparents die or your parents die. And so [death is] a taboo topic just from the way our culture treats it and [the fact] that we don't have daily experience with it. I would not wish upon any culture that they face death all the time, but it does change one's notion of how one treats death, how one sees it, how one anticipates it. I think it's taboo in part because the larger culture treats it as taboo. Some people are comforted by more conventional religious points of view. There's an afterlife they can imagine. I think other cultures have come to a more relaxed view of death.

Lee Lipsenthal, who had several months to live at the time of this interview, expressed his frustration about our collective worldview:

I think the structure of our society right now is one of a fear of death. I hate to sound so blunt about this, but it's a whole anti-aging movement. You're a loser if you die. You're a loser if you get old. And our society has set it up so that death and aging are the enemy, whereas they are inevitable.

## THE DENIAL OF DEATH

Ernest Becker spent a significant portion of his career seeking to understand the fear of death. While he was a professor of anthropology at the University of California, Berkeley, he published his seminal and Pulitzer Prize–winning work, *The Denial of Death*.[1] With this book, he awakened a conversation in American society about the cultural meanings of death. Becker saw the endemic denial of death as a source of pathology in our modern world.

"In our culture we have done a tremendous amount to deny our own mortality and in that process we have been initiated into a kind of pathological social organization," Becker wrote.[2] He saw the fundamental motivation for human behavior as a biological need to control our basic anxiety about death. "This is the terror to have emerged from nothing, to have a name, consciousness of self, deep inner feelings, an excruciating inner yearning for life and self-expression—and with all this yet to die."

Becker's work and theory about death denial has catalyzed decades of research on what is called *terror management theory* (TMT). To learn more about this provocative theory, I interviewed Jeff Greenberg, who, along with fellow social psychologists Sheldon Solomon and Tom Pyszczynski, developed TMT.

Inspired by Becker, Greenberg and his colleagues proposed that a basic psychological conflict results from having a desire to live but realizing that death is inevitable. This conflict produces terror, and this terror is believed to be unique to human beings. Each of us, according to TMT, is holding this suppressed death terror, largely at an unconscious level. In other words, we are not aware of the fear. To buffer ourselves against death terror, we seek to boost our self-esteem by affiliating with cultures or groups whose values provide our lives with meaning. For example, religious affiliation is a strong factor in Becker's model (for example, if you are Christian, or Hindu, or Jewish, you will affiliate with people of the same faith tradition). As we are confronted with our own mortality—or, in psychological terms, as our *mortality salience* increases—the theory predicts that we may become more aggressive and violent to other groups that hold different opinions, values, and worldviews than our own. At the same time, we may identify even more strongly with our "in-group," which offers us a greater sense of social support and security against outside threats.

The theory can be broken down into three main hypotheses. The first is the *mortality-salience hypothesis*. This asserts that an awareness of death leads people to defend or uphold their worldviews and seek self-esteem. In short, we want to feel good about ourselves. The second is the *anxiety-buffer hypothesis*. Here the idea is that high self-esteem, secure relational attachments, and deep religious faith should buffer people against death-related thoughts. The third, the *death-thought accessibility hypothesis*, proposes that when qualities such as self-esteem are undermined, we may experience increased vulnerability to death-related thoughts and maladaptive behaviors.

Testing these hypotheses has led to a series of novel experiments that simulate the real world under controlled research conditions. In one experiment, the team of social psychologists tested the mortality-salience hypothesis by working with municipal court judges who, by profession, are charged with upholding the dominant social worldview defined by the law and a sense of fairness.[3] To test the hypothesis, they asked the judges to set bonds for alleged prostitutes. Before the judges did so, half of them were given mortality-salience prompts. The researchers asked the judges two questions: "What emotions does the thought of your own death arouse in you?" and "What do you think will happen to you as you physically die and once you are physically dead?" The researchers then measured the amount of bail that was set by the judges who had been asked to think about death against the bail set by the other judges. They found that mortality salience had a very large effect. The judges who had been asked the mortality-salience questions set the bonds at an average of about four hundred and fifty dollars. The other judges set bonds of fifty dollars.

Since this original study, there have been hundreds of other studies that have replicated the results using different measures and in different countries, confirming that people became more rigid about their values when threatened by death awareness. According to Greenberg, "when we are reminded of death and it's sort of close to consciousness, we grab on more tightly to the structures that protect us from death. Those structures are a worldview that imbues life with meaning and a sense of permanence, and the self-worth that we derive from that worldview." In the case of the judges, this self-worth involved upholding the law as they understood it. Their conviction about the values they protect was amplified in response to the death reminders that half were given.

To test the second TMT hypothesis, the anxiety-buffer hypothesis, the researchers created a seven-minute video that highlighted images of death.[4] They also created a control video that consisted of neutral images unrelated to death. Prior to showing the videos to study volunteers, the researchers gave them false personality feedback. (With this feedback, researchers took advantage of what psychologists call the Barnum effect, named after famed showman P. T. Barnum: when people read flattering descriptions of themselves, they become gullible.) Based on questionnaires that the volunteers had filled out when they were recruited for the study, the experimental subjects were told either that they had a lot of potential for creativity and would achieve all their goals, or that they were just okay. After showing the subjects the death-image and control videos, the researchers measured the subjects' self-reported anxiety levels. When people "felt really good about themselves, they could watch those death images and not get anxious."[5] These anxiety levels were compared within the subjects who looked at both videos.

In another study, the social scientists introduced what they called the *hot-sauce paradigm*. They organized their test subjects into two preselected groups: liberal and conservative. They put each subject in a room and told them that a second person, seated in the other room, was either liberal or conservative. They then told the subjects to administer hot sauce to the second person as a punishment. When the researchers manipulated the subjects' mortality salience beforehand, the subjects administered significantly more hot sauce to those people who disagreed with their political views.[6]

Researchers have also explored mortality salience in the context of various forms of extreme worldviews. In one study, the researchers explored Islamic extremism. Student volunteers in Iran were either given a mortality-salience prompt or not. They were then instructed to read an interview on martyrdom or an interview on peaceful solutions to a conflict. The volunteers who received the mortality-salience prompt were significantly more inclined to think favorably of martyrdom than the control group (those who didn't receive the prompt).[7] A similar study focused on politically conservative Americans and found that the group in the mortality-salience condition advocated for more violent measures, when dealing with foreign conflict, than the control group did.[8]

Becker described how our death terror ultimately can lead us to a perception that the world is frightening. There are various ways in which we attempt to manage this terror. Becker's model says that we conspire to keep our terror of death unconscious by pretending that the world is manageable, that humans can have godlike qualities, and that the self is immortal. Society reinforces the creation of hero systems that lead us to believe that we transcend death by participating in something of lasting worth—the pyramids, for example, or great cathedrals, or an orbiting space station, or the Internet. To deal with our own mortality, we feel compelled to create works or take actions that will live on after we die, giving us a perception of immortality.

In the foreword to *Denial of Death,* writer and philosopher Sam Kean notes, "We achieve ersatz immortality by sacrificing ourselves to conquer an empire, to build a temple, to write a book, to establish a family, to accumulate a fortune, to further progress and prosperity, to create an information society and global free market. Since the main task of human life is to become heroic and transcend death, every culture must provide its members with an intricate symbolic system that is covertly religious."[9]

But in the process of developing the hero myth, as comparative religious scholar Joseph Campbell called it, we have created personal and collective struggles. As our emotional stability is threatened, Kean explained to me during an interview, the existence of alternative worldviews can cause us to question our own convictions and beliefs. As we saw from the TMT research, this may lead us to feel defensive and hostile toward people who are different from ourselves—what sociologists call the *out-group*. And this defensiveness can lead to conflicts, including religious wars, state conflicts, and racial battles. In Kean's words, we suffer from a crisis of heroism: "Our heroic projects that are aimed at destroying evil have the paradoxical effect of bringing more evil into the world. Human conflicts are life-and-death struggles—my gods against your gods, my immortality project against your immortality project."

## GAINING PERSPECTIVE ON DEATH

The studies of TMT show that reminders of death can be both personally unsettling and socially disruptive. And yet there appear to be ways in which our awareness of death can lead to positive psychological and behavioral outcomes. Addressing the third hypothesis, the death-thought accessibility hypothesis, an international group of social psychologists, led by Kenneth Vail at the University of Missouri, reported in the journal *Personality and Social Psychology Review* that there is ample evidence to suggest that "the management of death concerns can play a key role in motivating people to stay true to their virtues, to build loving relationships, and to grow in fulfilling ways." This may include a renewed commitment to exercise or a healthy lifestyle.[10]

Other research indicates that there may be anxiety buffers that lead to positive and life-affirming behaviors. Linking death awareness to values of tolerance, appreciation, and curiosity about alternative worldviews can help people feel less threatened in the face of mortality-salience prompts. This move from "me" to "we" is predicted in the worldview transformation model. Direct personal experiences, such as near-death experiences, have been shown to enhance intrinsic values, such as love and compassion. This is compared to extrinsic goals, such as attaining wealth or success. In this way, these experiences appear to promote prosocial behaviors that enhance individual and collective wellbeing.[11]

Given how strongly the fear of death affects our beliefs and behaviors, several colleagues and I wanted to research ways of transforming fear, so that people can live a fuller life. To do so, we first created an online course that reviewed diverse worldviews concerning death and the afterlife.[12] The course encouraged group discussions and sharing direct personal experiences related to death. We used excerpts from interviews with individuals representing different cultural, spiritual, and religious traditions. This provided the course participants with an opportunity to explore and develop appreciation for diverse worldviews.

After the course, my colleagues and I tested the impact it had had on people's worldviews of death and the afterlife. To do so, we studied data from journals and questionnaires that we had sent participants before the training and again after the training. We gave participants two mortality-salience prompts that had been used in previous research—"What do you think will happen to your body when you die?" and "What are the emotions that the thought of your own death arouses in you?"—and asked them to answer these questions in their journals. By analyzing the language participants used in their journals before the course and after, we found that their views about death had changed significantly by the end of the course. Participants were less centered on their body and less focused on death, and they were less likely to use first-person pronouns in their journal writing, suggesting a reduction in their personal identification with death. They also appeared to demonstrate more insight about themselves, and their writing included fewer negative references to death as something to be feared and denied.

For example, before the course, one participant responded to the question "What are the emotions that the thought of your own death arouses in you?" by writing, "Fear of having regrets. Almost panic at the thought of leaving dreams and talents unfulfilled or wasted." After the course, the same person wrote, "My spirit will be surrounded by an atmosphere of love and peace, and I will feel no real regret at what I am leaving behind." This person experienced an overall emotional shift. The course seemed to have given her a sense of cosmic unity and a readiness for her own death.

Another participant expressed a similar shift in worldview. Before the course, he noted, "The most prominent emotion is worry that I have not accomplished all that I have set out to do. What if I haven't done enough or have somehow been less than I was supposed to be or experience?" After the course, he wrote, "I feel excited by the prospects that I have completed my life journey and that I will be moving on to new adventures."

Follow-up comments told us that the course had deepened some people's perspectives about death and invited open-minded curiosity about and exploration of diverse worldviews on the afterlife. One participant noted:

> Besides the fact that my beliefs feel strengthened, I have a better understanding about beliefs of other cultures and feel that I have added certain pieces to my own beliefs. . . . In particular I want to strive to approach other worldviews and my own with respect and humility. I want to continue to allow openness and space for something different to enter.

Another participant said:

> My personal experiences and beliefs are real, and other people's are real, too. It's okay to not "know" for sure what the afterlife is like, but by sharing our views and experiences, we open the conversation in ways that still support the existence of an afterlife, regardless of variation of perspectives.

This research suggests that raising death awareness in a supportive and engaging environment can help ease people's fears and reduce their resistance to worldviews that are outside their own, thus counteracting the cultural pathology that Becker wrote about so powerfully.

## SHIFTING THE FEAR OF DEATH

The worldview transformation model predicts that engaging with worldviews beyond our own is one way of confronting death in order to transform our fear of it. Other people have faced their fear of death and shifted it to life-affirming values through their direct personal experience, their spiritual beliefs and practices, and their philosophy of life.

### Embracing Death to Embrace Life

Dean Ornish is a renowned physician, a clinical professor of medicine at the University of California, San Francisco, and the founder and president of the nonprofit Preventive Medicine Research Institute. His personal story demonstrates how embracing death, rather than avoiding and fearing it, frees us to live better lives. During an interview, he told me how, when he was in college, he came very close to committing suicide.

> I first got interested in doing this work really when I was an undergraduate. I was at a small, very competitive university where half the student body graduated first or second in their high school classes. It also turned out to be the school that had the highest suicide rate in the country per capita. . . .
>
> The more I worried about doing well, the harder it became to study. And the harder it became to study, the harder it became to do well. I got into this vicious cycle where I literally couldn't sleep for a week straight, and I got profoundly agitated. That kind of sleep deprivation alone is enough to make you a little crazy.
>
> So I remember sitting in physics class, and I thought, "I'll just kill myself! Why didn't I think of that earlier? That will put an end to all of this pain." . . . So I made plans to kill myself, and I was going to jump off a tower. [Then] I realized my parents would not be very happy. So I decided I'd get really

drunk and run my car into the side of a bridge, and everyone would think I had just gotten really drunk and I wasn't really trying to hurt myself. That would be easier. I was a little crazy, but it made perfect sense to me at the time.

This was in 1972. Meanwhile, my older sister had been studying yoga with a swami named Swami Satchidananda. He was an ecumenical teacher who had come to the US in the midsixties. . . . My parents decided to have a cocktail party for the swami—which in Dallas in 1972 was pretty weird. So he came into our home.

There's an old spiritual teaching that "when the student is ready the teacher appears," and that was certainly true for me. When the swami came into my parents' living room and said, "Nothing can bring you lasting happiness," it was really validating, because everybody else had said, "Of course, things will make you happy. Just do this and do that and don't do that, and then you'll be happy." I realized that that wasn't true, [and] here he was validating that . . . I thought I was ready to kill myself, and he was glowing, and I thought, "What's the disconnect here?"

This meeting with Swami Satchidananda was a moment that transformed Ornish's life. He realized that happiness and peace of mind come from within us, not from the external world. Such qualities are something we have already, and they are not something we can lose. He explained to me, "One of the great paradoxes of life [is that] we run after all these things that we think are going to make us happy and peaceful, and in the process we disturb what's there already."

Ornish also discovered that many spiritual practices aim to calm our bodies and minds in order to let us experience what we already have within. Compelled by these insights, Ornish decided to try meditation, acknowledging that he could always return to his initial game plan, suicide, if the spiritual practice did not work. In this meditation process, he began to get glimpses of inner peace and well-being. It was a gateway to personal transformation for the young man.

> Because I came so close to killing myself when I was in college, because I was so profoundly depressed, I naturally started to ask questions like: What is death about? Do you just close your eyes and go to sleep? Or is it something more than that? And I began to read up on this voraciously. I began having my own experiences. When you meditate enough, you realize that we have a body, but we're not our body. We even have a mind, but we're not our mind. There's something that survives death, that goes from one class to another, one body to another.
>
> The paradox, and I've seen that in my own personal life, as well as in the many, many people that I've worked with over the last thirty-five years, is that . . . until you fully embrace death, you can't really live fully. It's a cliché because it's true.

Just as he was ready to face death head-on, Ornish unexpectedly found himself embracing life. Today his work centers on helping people live in optimal health. The physician attributes his professional success to his own close encounter with death.

Ornish's experience follows that described by the worldview transformation model: he had a profound destabilization; he had bottomed out. He then discovered a gifted teacher and began his own spiritual practice, which grounded his work as a healthcare provider and scientist. He sought to understand what he was experiencing through meditation, and what he experienced helped him to understand death. Through his own worldview shift, he found the path to living deeply.

> Everything I've done in my professional life people thought was crazy because people thought it was impossible, but I would never have done it if I hadn't come so close to dying earlier. . . .

When you really are interested in learning
something, it tends to be more successful . . .
because you don't bring all that anxiety and fear.
You're willing to try things that other people would
just think are too risky because you've come so
close to death that you want to. When I decided to
not kill myself, I wanted to live as fully as possible
because I couldn't live half a life.

## Choosing Life

Like Ornish, Noah Levine was once suicidal, longing for death most
of his early life. He believed on some level that death would help
him escape suffering. His father had been active in hospice work, so
Levine had a familiar relationship with death. Then, over decades
of Buddhist practice, Levine's relationship to death changed "from
wanting out, to actually being quite happy to be in the body." Today,
the author of *Dharma Punx* and *Against the Stream* is also a Buddhist
practitioner and counselor. He teaches Buddhist meditation classes,
workshops, and retreats and leads groups in juvenile halls and pris-
ons on the role of mindfulness for living well. He has studied with
many prominent teachers in both the Theravada and Mahayana
Buddhist traditions.

In contemplating suicide, Levine never felt that death was "lights
out." Rather, he had an "understanding that death is just a transition
from one form of existence to another, without a lot of attachment
or fear about death—feeling like it is such a natural process." A core
tenet of his mindfulness involves facing death directly, acknowledging that our bodies die and are not who we really are. An important
ritual within many Buddhist practices involves visiting funeral
grounds to watch the bodies being cremated. He says, "Just as that
body is burning, so will this one eventually. So find a place that is
not this body to take refuge in a spiritual understanding." (We will
explore death-preparation practices in chapter 6.)

Today suicide is not part of the picture for Levine. His goal is to
live fully and with purpose. During an interview for the transformation study, Levine underscored the certainty of death coupled

with its uncertain timing. He explained that this two-fold nature of death gives him a sense of urgency that impacts how he spends his time and what he does with his life. He draws on a worldview that includes reincarnation. He explained that if you don't do your work, you return in another body, and another, until "you've done what needs to be done, which mostly is freeing oneself from delusions and greed and aversion and confusion." He continued:

> The fact that I reflect on death a lot does inspire me to [spiritual] practice. And it does influence my practice of saying, "I don't know how much time I have, so I better pay attention." Rather than my earlier lifetime where I was kind of like, "Sure, I'd love to trade in this existence for a different one." Now, it's like, "Oh no, this existence is what's a given. Here it is. I don't want to trade it in for another one. Actually I'd like to get free, and not, you know, keep doing this cycle. I'd like to have the kind of freeing experience that the Buddha talks about—of entering the deathless and not continuing to take rebirth.

## Redefining Identity

Many spiritual teachers believe that we can shift our views of who we are and that doing so offers a portal to worldview transformation. An aspect of death that causes fear is the question of personal identity: Who or what dies? Answering that question of personhood can help us to reformulate our relationship with death, says Satish Kumar. A former monk, longtime peace and environmental activist, and the editor of *Resurgence & Ecologist* magazine, Kumar explained to me his worldview:

> My identity, what we call small identity—like my label, my name, my nationality, my religion—doesn't survive bodily death. These are small identities. If I am a member of the universe and a member of the earth community, and I am part and parcel of the life

force, that is my identity. It is my true identity, or my
primary identity.

My secondary identities are that I am an Indian.
I am a certain age. I was born in the Jain religion,
et cetera. All these are secondary identities. We
need not be afraid of losing secondary identities.

The worldview transformation model tells us that positive transformation leads us to shift from the small "me" to an expanded "we"
that connects us to something greater than ourselves. Kumar echoed
this concept:

Our primary identity is that we are members of the
life force. That life force continues. It's eternal. It's
infinite. It's dynamic.

If we become static in one body, and never die,
and are afraid of dying, that means we are blocking
the dynamic force that is ever changing. If we block
that, the world will be a boring place.

Like Becker, Kean, Greenberg, and others, Kumar argues that people's fear of death leads to maladaptive behaviors. It can block the
process of positive worldview transformation. For him, the key to
overcoming our fear of death is a worldview that heals our separation
from the natural world. The ecologist describes the shift from external motivations to intrinsic motivations that may help us to identify
our true nature. Expanding our sense of self and who we really are
can help connect us to a broader and more transcendent reality.

We want to live forever, and because we want to live
forever, we want to own, we want to possess, we want
to control. Therefore, we want to own nature, we want
to own the land, we want to own property, we want
to own the people, and we want to own relationships.
That is what destroys the normal functioning of the
cosmos. Earth is a part of that cosmos. . . .

If we expand our consciousness, expand our mind, go out of this body, and understand that the whole of earth is my home, and I am an organ of the whole earth body, and that earth body is a part of the whole of cosmos, then my mind is expanding. And then, as vast mind, I touch the mind of God. That is how we can liberate ourselves and not possess this land, or this relationship, or this house, or this money. Me, me, me, and I, I, I—this kind of possessiveness will melt away. And that will be the true, deep ecology.

The greatest friend of fear is ignorance because ignorance leads to fear. . . . Remove that ignorance and you become aware that we are part of the continuum, a part of the evolving. The moment you realize that, then you are not afraid.

## Embracing Our Gifts

Michael Bernard Beckwith is the founder and spiritual director of the Agape International Spiritual Center, headquartered in Los Angeles. A multicultural, transdenominational community in the New Thought–Ageless Wisdom tradition of spirituality, Agape has thousands of local members and global live streamers. The vision and mission of Agape and its ministries are grounded in the principle of being a compassionate, beneficial presence on the planet.

Beckwith shared his philosophy on how to skillfully work with the fear of death:

As we begin to birth spiritual insight, we grow in our understanding that all beings are one, that all existence is cosmically interconnected. We further realize that we possess more than enough of all that is good and perfect, including our nature of eternality. We never die, because we've never been born. We're living our human incarnation; but as spiritual beings, we are part of an evolutionary continuum taking place in many dimensions.

When individuals catch that life on earth is impermanent, that their human incarnation and delivering their gifts and talents will come to a closure at some unknown time, this causes a high level of anxiety. The human ego—the sense of being a self, separate from the whole—becomes very anxious about when its time will be up and makes life choices based on actuarial tables created to predict a person's lifespan by age, as though this were the main criterion.

How much wiser it is to dismiss timeframes and make choices and decisions from a consciousness of, "What gifts and talents am I to cultivate and deliver before I leave this three-dimensional realm? While I'm here, how can I be a beneficial presence on the planet?" This form of self-inquiry vibrates at a much higher frequency than, "I'm afraid to die!"

## Being in the Present

Living in the moment can offer salve for the death terror that many people experience. Brother David Steindl-Rast, whom we met in chapter 1, explained to me:

> People who are dying are forced to be in the present moment because they have no more future. The closer they come, the more they have to be in the present moment. One of the aspects of Being that [psychologist Abraham] Maslow identified in the peak experience was beauty, because we are in direct contact—for a split second our little ego drops, and we forget about it—and we are in direct contact with Being.
>
> To simply be present where you are allows the transforming power of the universe to transform you. You don't have to do anything to grow older. That happens by itself. You don't have to do anything to digest your food. Nature does that, and you couldn't

do it if you wanted. So you don't really have to do anything to grow spiritually and to transform spiritually. It happens when you are not getting in the way.

What I'm saying is that when you live in the present moment, you are touching upon something that isn't subject to time. We live in what T. S. Eliot calls "the moments in and out of time." We know what now means, and now is not in time. It's a little strip between the past and the future. The now is beyond time.

Lee Lipsenthal also valued living in the moment, embracing his most authentic self. He told me he felt no fear around his impending death because his way of dying wasn't any different from the way he lived.

The beauty of what's already been is enough. People feel that I'm brave for talking about what it's like to be dying in a public venue. I don't look at it as bravery; I just look at it as me, dying. This is what I do. I teach. I'm out there being with people, enjoying life, enjoying play. It's who I am. I just happen to be dying now. This thing that is looked at as bravery or resiliency is just me dying. The "me" hasn't changed. The physical entity is changing for sure; I've got lumps in my neck. But the core "me" hasn't changed. And so this bravery resilience thing is to me, just me being with cancer. I wouldn't know how to do it in another way.

## Finding Unconditional Love

Tony Redhouse is a Native American sound healer, spiritual teacher, eagle/hoop dancer, and award-winning recording artist. The creator of Native American Yoga, he is a traditional Native American practitioner and consultant to Native American communities and behavioral health organizations, where he teaches seminars on Native American culture. He is also a cancer survivor. For Redhouse, unconditional love—love without judgment—is the greatest buffer against fear, including the fear of death.

> The only reason we have fear is because of judgment.
> . . . In our mind, there's some type of judgment
> because we have not met certain criteria. We're
> comparing our self with some ideal. Removing that
> fear, any fear, and especially the fear of death, is being
> able to understand unconditional love. It's being
> able to embrace that. And that goes back to our self.
> Somebody said, love your neighbor as yourself. When
> I was going through medical treatment for forty-eight
> weeks, which involved some chemo, I looked in the
> mirror, and I was angry. I was upset. I was depressed.
> My energy level was down, I just wanted to lay under
> the covers and go to sleep all the time.
>
> During that time, I finally got up, looked in the
> mirror and looked into my eyes for the longest time.
> And I said, "Tony, I love you just the way you are.
> Everything you've been through, every relationship,
> every failure, every heartache, every celebration, every
> success, I love you just the way you are. Exactly the
> way you are. You don't have to change one thing. I
> embrace you unconditionally."
>
> When we can find that unconditional love that
> does not have any judgment, when we can embrace
> everything that we are in one moment, even for one
> second, look in a mirror, and say, "I love you just
> the way you are. Even with everything that you've
> been through in your life—divorce, addictions,
> everything—I love you completely," that wipes out
> the fear of death.

## Caring Community

For physician Gerald Jampolsky, a key to healing our fear of death is love and caring. Loneliness and social isolation can lead to early deaths, and holding on to old emotional baggage may lead to suffering. Jampolsky offers Attitudinal Healing, which he explains is "a cross-cultural method of healing that helps remove self-imposed blocks such as judgment, blame, shame and self-condemnation that are in the way of experiencing lasting love, peace, and happiness."[13]

People who have a life-threatening illness may be panicky and afraid of dying. This suffering can be transformed, says Jampolsky, as these people participate in a community that fosters love and forgiveness:

> They're in a group of similar people where they're giving help as well as receiving it, and finding out "the more I give my love, the more I'm able to stay in the present."
>
> They're getting out of the old paradigm that the past is going to predict the future. They're learning to live in the moment. They're learning to forgive, because when we don't forgive, it creates toxins in our bodies that cause us to hurt ourselves. So we hold on to anger around someone else, and oftentimes even in the dying process medications like morphine may not be useful because the pain is still there. But if [they] open up the possibility of looking at some places where they haven't forgiven themselves or others, all of a sudden the medicine starts to work. . . .
>
> The purpose of our group is to practice forgiveness, not making judgments, and giving. As we give, we receive. The benefits come from really staying in the present, not asking questions that will bring about fear about what's going to happen tomorrow, or what happens when the doctor finds my x-ray has gotten worse. Instead, [we focus on] . . . realizing that when you're in a hospital bed and people are coming to see you, a lot of them are afraid and fearful of saying the wrong thing.
>
> A lot of people may not be coming around to see you, and you wonder if you're being rejected, when

really they are fearful. Rather than getting upset and angry that old friends aren't coming to see you, you send them love and begin to feel a peace and a joining, not a separation. We learn that the purpose of relationships is joining, not separation. So it's a whole new way of living life.

## GLEANINGS

Terror management theory holds the critical assumption that our fear of death lies buried beneath the threshold of conscious awareness. Both the terror of death and our heroic impulses to overcome death are linked to our worldviews. Bringing death and how we avoid our fear of it into conscious awareness can deepen our self-inquiry. As we begin to bring death out of the realm of the unspoken, we may better integrate our understanding into our everyday experience. Staying in the present helps us to be fearless. Without striving, we may get out of our own way as we face the inevitable.

Data from the experiments in social psychology and the worldview transformation model suggest that death awareness can motivate people to reprioritize their life goals and expand their self-worth and identity, prevent harm to others, and promote social harmony. By exploring our own and others' worldviews about death, we may become more creative, innovative, and flexible about our relationships with ourselves, our community, and the world in which we live. Exploring worldviews about death becomes the base from which we can reflect on our own potential resistance to fundamental change and becomes the starting point for worldview transformation.

If we treat death as a great mystery, we may see it as an adventure and an opportunity to engage the unknown. If instead of denying death, we create a positive awareness around it, we can open to a new understanding of ourselves and others. Learning and appreciating the depths of human experience can be a doorway into our own evolution. In the next chapter, we will explore the nature of direct personal experiences of death and near-death and consider their role as catalysts for deep and lasting growth and wellbeing.

## Just Being

Science shows us that self-esteem can be a powerful buffer against the fear of death. In order to invite this tool into your life, begin with this simple exercise.

Sit quietly. Bring a small smile to your face. Feel the muscles in your cheeks as you hold a positive intention.

As you smile, bring to mind a positive quality or characteristic about yourself (for example, "I'm creative," "I'm funny," "I'm a good hugger," "I'm smart"). Begin to breathe into this thought about your positive quality. Continue to smile as you enjoy this positive aspect of who you are and how you feel about yourself.

Take a few moments to absorb this experience and feel the goodness that washes over you. Then use your journal to record any feelings, sensations, or personal insights that arose for you.

If it feels right, share what you have discovered with a family member or friend, inviting them to reflect on their own positive qualities. In this way, you may together begin to form a caring community that can enhance your self-esteem and help transform any negative reactions you have about death.

# Chapter 3

# GLIMPSES BEYOND DEATH AND THE PHYSICAL WORLD

Noetic experiences . . . states of insight, unplumbed by the discursive intellect, all inarticulate though they remain, and yet they carry with them a curious sense of authority.

WILLIAM JAMES

Simon Lewis was a rising film producer. With the production of *Look Who's Talking* under his belt, he had set up a deal for Nick Nolte's next picture, and his film *Old Friends,* produced for HBO, had won some Emmys. And then, he was forced to change his life.

> I was on my way for dinner one evening, and a big van traveling at seventy to eighty miles an hour ran a stop sign and bulldozed my car against the curb. My car took off and hit a tree in midair. This was less than a mile from Cedars-Sinai Medical Center, one of the biggest hospitals in LA. Some paramedics were off duty on their way home from work when they saw the crash. They . . . told the police that there were no survivors in the car and that I was dead.
>
> This place of pure consciousness that I was experiencing was actually inside the lowest level that is measured of coma, which is called a Glasgow Coma Scale 3. I was there for over a month, not expected to live. . . .

When I came out of the coma, I recognized my family, but remembered none of my past. In fact, it was only after a few weeks that pieces of memory started to come back.

A memory from my coma came to me of a protector who was traveling with me through this inner space, through this journey into infinity. And very slowly, in the middle of the night, it occurred to me who that protector was. I realized it was my wife. I was very pleased, and I told the night nurse, who came to turn me over every twenty minutes, the good news.

The following morning, my mother came to the hospital to explain that I hadn't always been in this room and in this bed. The reason I was in this room and this bed was because I'd been in an accident. Before this room and this bed, I'd been a filmmaker, that I was married to Marcy, and that she had died instantly in the crash. During my month in a coma, she had been laid to rest in Phoenix, Arizona.

So I lost everything that really mattered most to me. And that started this journey through consciousness to find what makes us who we are, to see if I could recover who I was, and also to find a reason to go forward.

Lewis had had what William James called a *noetic experience*—a direct, personal encounter with death and a world beyond the physical—that transformed his life. I have been chasing these unplumbed sources of illumination for many years through my life and work. Over and over, I have found clear support for the value of these experiences in shaping and transforming our worldviews in positive, life-affirming ways.

Noetic experiences move people from a strictly materialistic view of their existence to one that has ineffable, spiritual, or mystical dimensions. Many people report rich and compelling states of consciousness that shape their understanding of death—and what may lie beyond. People's encounters with the transcendent offer a personal kind of methodology for understanding the transformative potential of their own inner experiences. These encounters are a way of knowing about reality that transcends logic and reason. In the context of death and dying, such experiences may be painful, rocking people from their natural steady state. Noetic encounters with death can also offer inspiration, hope, and mindfulness in the face of life's ultimate transformation.

## A NOETIC JOURNEY

Lewis had a profound noetic experience that included alternative states of awareness. Even though he never wanted the experience, he was able to use the memory of his time in deep coma to transform his sense of embodiment and his understanding of what is true for him.

As Lewis recounted his experience, he raised an issue of epistemology, or how we know what we know. He didn't recognize his wife as his wife. Rather, she was more of a presence that guided him back to a waking state.

> I was traveling through an ancient grove, on a boat, and I could hear the sounds of rain pattering down on the cabin. There was a person who was up on the deck, with whom I knew I would be safe if I could only just go up to be with her. But I didn't know definably who she was. It was my sense memory at the lowest level of coma [that] I call inner space.
>
> I described everything that occurred to me there as both fresh and familiar, which seems impossible. Something can't be both fresh and familiar. But it is, inside the deepest level of consciousness. At that deep level, everything that you're experiencing is being created at a deeper level still of consciousness that is extending way down into infinity. You can't normally see down the slope of consciousness yourself because, by definition,

you're looking at it from a higher level of consciousness. You can't see your own subconscious normally.

Lewis's worldview had shifted, and he opened to a new ontology or understanding of what is real and meaningful for him. He experienced a profound transformation through his direct personal experience. His words suggest that not only are there different ways of knowing and experiencing reality, but, in fact, there also may be different realities that coexist. He elaborated for me:

> How did that change my worldview? There are
> different levels. From race memory that transmits
> through the generations, and from inner space, I
> gradually came up through these infinitesimal
> steps—they are very, very gentle gradients—through
> flat time that led me to see consciousness as one
> continuum. I'll never forget when I saw the river of
> time; that certainly has changed my life. I believe that
> when I saw my own subconscious, I saw time stop.
> And it's always stayed with me because I've realized
> that my subconscious doesn't know the day. It knows
> usually whether it's day or night, but it doesn't know
> the time or the date. It just knows the infinite now
> in which consciousness lives. That has completely
> changed my life . . .
>
> When we die, the slopes continue. I continued,
> and I realized that the mind is always curious. Where
> that curiosity goes at the point that the body dies . . .
> is unknowable. What I was experiencing in a state
> of consciousness at a Glasgow Coma Scale 3 was a
> constant voyage of vistas that kept opening, one after
> the other.

Lewis's process of personal discovery showed him that the realness of reality is impacted by the experiences that we have and the view we hold. As William James implied more than a hundred years ago, noetic experiences can be realer than real. There are striking accounts of these experiences that suggest that our current perception of the world is only a limited subset of what appears to be possible. Like Lewis, many people have reported near-death experiences that speak to a world that lies beyond the physical and material, and they have been personally transformed by these experiences. Opening to a perspective that invites in our direct knowing and subjective experience is core to worldview transformation.

## CATALYSTS FOR TRANSFORMATION

Many things inform what we hold to be true in our lives. For example, sensory empiricism is the method of understanding objective reality that grounds modern science. From this perspective, what is true is what we can experience and measure with our five senses. From this lens of perception, we are able to define reality based on its physical nature; we are real in terms of our own embodiment. But other dimensions can be seen as unreal by or are excluded within the scientific worldview. Traditional science and medicine typically treat mystical states and contemplative or intuitive knowing as the result of delusion or pathology. Yet insights and experiences that transcend easy physical explanations play a powerful role in how people understand death and what may lie beyond.

The worldview transformation model tells us that such ineffable insights can catalyze shifts in our perceptions of reality. It is also possible that such experiences can remain unexplored or ignored, depending on a person's life circumstances, awareness, or cognitive defense mechanisms. Bringing intention and awareness to our worldview around death can turn a painful experience into a gift. It may lead to alterations in our goals, values, and priorities.

As we learned from the research on terror management theory (chapter 2), our awareness of death can lead to existential terror. It can also, under the right conditions, lead to life-enhancing choices and experiences. (We will explore some of these conditions in the chapters that follow.) Simon Lewis's experience in the depths of coma, for example, led him to restructure his own worldview. The existentially charged experience offered him the opportunity to restore and refashion his perception of himself and his place in the world. More importantly, the changes in his life have been long-term. His experience led to an identity shift that presented him with a feeling of interconnectedness and unity. His worldview expanded in the wake of crisis. He has also been moved, as predicted in the worldview transformation model, to use his experience to help other people transcend their fear of death.

## CONNECTIONS BEYOND DEATH

Jean Watson is a distinguished professor and dean emerita of the University of Colorado College of Nursing in Denver and founder of the Watson Caring Science Institute. She had her own noetic experience, involving the death of her father, when she was sixteen.

> He died abruptly of a heart attack. I was at high school, so I missed his death. When I went home, he wasn't there. He was dead—and he'd been taken away. I kept having this fixation about wanting to see him and saying good-bye to him. I remember it being so unfinished. All I wanted was to see my father again. Of course, I did at the funeral, but it wasn't the same. I just had this desire to see him again.
>
> The very night of his funeral, he came to me in my bedroom. It scared me as a young girl. I remember looking up and seeing him, fully seeing him as an apparition in the doorway of my bedroom. I was startled. I closed my eyes, ducked under the covers, looked up again, and he was still there. It was scary and also comforting, as if he had come to let me know that it was okay.

Because she was frightened by her vision of her father, Watson sought out a trusted mentor. She asked her biology teacher if he thought it was possible to see people after they die. As she recalled,

"He told me we could. I was so relieved." Her teacher helped Watson understand and integrate her noetic experience into her existing worldview. The experience helped her resolve her feeling of disconnection from her father and offered her comfort.

As a young nursing student studying biology, Watson came across the idea that energy can never be created nor destroyed. This concept helped her understand the nature of death, and with it, she began to reformulate her worldview. She shared with me the transformation that came from this conceptual leap:

> If somebody dies, energy, their soul, is still out there somewhere in the universe. That was a very formative experience for me personally, in terms of being on this side of the veil, rather than the other.
>
> More recently, my husband died abruptly of suicide. That was very disarming, to say the least. A similar phenomenon [to that with my father] happened with my grandson as we were sitting in the memorial service. He kept jerking his head about. I thought he was embarrassed because I had my arm around him. But he came to me after the funeral and said he wanted to talk to me, and he didn't want anyone else to hear. So he went to my room with me, and he told me that when the minister started speaking, he saw his grandfather on the ceiling of the building. He described him as a constellation and [said] that he was trying to tell him something. He kept jerking his head back because he wasn't sure he was seeing what he was seeing. It was the same experience that I'd had as a teenager.

Trying to make sense of such noetic experiences can be challenging when they fall outside the standard worldview of our dominant Western culture. In fact, such experiences have often been pathologized by the mental health community, which may fail to see their transformative potentials. For some people, learning to accommodate such information into a new worldview can help them refashion their understanding of what constitutes a meaningful and significant life. In this way, their fear of death may be attenuated. Watson helped her grandson, just as her biology teacher had helped her years before. Later, Watson found another teacher who helped her with her own loss of her husband.

> The only sense-making I had of those experiences was when I was at a workshop with a Native American elder. I asked him about it. He said that we all come from the spirit world. We're here for a given period of time on the soul's earth plane to fulfill a soul's journey. When that soul's mission is completed, we return home to the spirit world.
>
> Our purpose here might be very grand, or it might be very small. It might be as simple as being at a certain place to help an old lady across the street. If that's when our soul's journey is complete, that's when we go home. I was trying to be discreet at that time, and I said to him, "Well, what if that soul goes home prematurely?' And he said, "That's your word, 'prematurely.'" He said that we're all dying. We're all choosing when we go; we choose when we come; we choose when we return. In his language they do not have a word for suicide.

These experiences over several decades have informed Watson's self-understanding. They have led her to shift her values and her beliefs about the world. They both challenged her worldview and led her to expand what she knows to be true about life and the possible survival of spirit following bodily death. While they were traumatic at the time, integrating these noetic experiences into her personal meaning system expanded her sense of purpose and possibility.

> All of those experiences, from adolescence to more recently with my husband, have informed me. They

also maybe opened up my third eye to make me more open to the other side and to the thinning of the veil—to acknowledge that there's just so much we don't know that we can be open to.

## THE TRANSFORMATIVE POTENTIAL OF NEAR-DEATH EXPERIENCES

Noetic experiences can occur with the death of someone we know, as in Jean Watson's case. Another type of noetic experience is the *near-death experience,* in which we ourselves die or come close to dying, only to return to life.

### Finding a New Place in the Universe

Joseph McMoneagle is not a person you would associate with mysticism. He is an ex–US Army intelligence officer who served in Vietnam and held a top-level security assignment with US Army Intelligence and Security Command. Later, after his noetic experiences, he was recruited to participate in the highly classified psychic-intelligence project now known as the Stargate Project.[1] Prior to his work on the Stargate Project, he had two near-death experiences that shaped his worldview in life-affirming ways. In the first case, he was in Austria, having dinner with his wife and a friend. Starting to feel sick, he excused himself to go outside.

> When I exited through the front door of the restaurant, there was a pop, like someone snapping their fingers, and I found myself standing on a cobblestone road. It was raining, and the rain was passing through my palms. I thought, "This is very peculiar." I looked over, and a body was half in and half out of this swinging door of the restaurant.
>
> I noticed that it looked very much like my body. My friend that had brought my wife to the restaurant had come outside and pulled the body into his lap. He was striking the body on the chest with his fist;

they didn't know what CPR was in 1970. I found out later that I had gone into convulsions, collapsed, and had swallowed my tongue. His solution was to keep hitting me on the chest with his fist.

They loaded the body into a car and rushed me to a hospital in Passau, Germany. That took quite a while since it was about sixty kilometers away. By the time they got me to the hospital, I had not breathed for a while, and my heart had stopped. And I was watching them—I floated alongside the car. I watched them cut the clothing off in the emergency room and stick needles in my chest. I had drifted up to the ceiling in the out-of-body state. I felt heat on the back of my neck and thought it was those bright lights near the ceiling. I turned to look at the lights and fell over backwards into a tunnel, accelerated through the tunnel, and when I came out at the end, I was enveloped in this very warm, bright light.

What McMoneagle reported is a classic near-death experience. While distressing for those on the outside, the event opened him to profound spiritual insights.

Instantly, I knew all of the answers to the universe. I knew that I was in the presence of God because that's what it had to be. I was overwhelmed with love and peace. Then a voice said, "You can't stay. You have to go back." I argued with it and said, "Nah, I'm not going anywhere." And then there was another pop, snap, I sat up, and I was under a sheet, naked.

I looked around, and there was a German lying in a bed next to me. I had been comatose about twenty-five hours. I was very excited and started telling him, "God's a white light. You can't die." He ran out and got the doctor who came in and sedated me.

I woke up a little later, and they were taking me to Munich to put me in a rest home where they would start doing brain studies. They were sure that I was crazy and that I'd suffered brain damage from the lack of oxygen. Over a two-week period, they were able to figure out that I'd not suffered any brain damage.

However, I was unable to reconcile the events. I was having spontaneous out-of-body experiences. I was hearing conversations going on four rooms away. I was reading people's minds that were walking into the room. I was psychically scattered. And I'd totally lost my fear of death—which the military noticed.

Eventually they let me out of the rest home, and I got to spend seven more years overseas doing some very strange jobs because I had no fear of death. For a long time, until 1985, I believed the white light is God, and that you can't die, that you survive death.

McMoneagle had a second near-death experience in 1985, when he experienced a massive coronary. While dying, he planned to exit his body and go toward the light that he had previously experienced. "I wanted to consciously experience death—[experience it] in a very conscious way," he explained. "That was thwarted by doctors."

In the dying process, I was able to see the light, but not go to it. For some interesting reason, I can't explain why, I was just not allowed to do it. But I could see the light, and I could see the light had edges. That created a huge philosophical problem for me because my definition of God is that God is an unlimited being, and an unlimited being can't have edges.

Following his second noetic experience, McMoneagle began to explore the intricacies of his worldview. As is the case for many people who have begun the transformative journey, he considered his experiences for a long time. After more than a year, he concluded that the light is what we are when we cease to be physical. That conclusion led him to a new understanding of time and space.

I think we become, in a sense, an almost pure form of energy. And in this pure state of energy, we coalesce into all of the knowledge that we've collected in all of our forms, many of the lives that we've lived.

I believe in multiple lives, not recurrent lives. I don't believe that we are born into lives in a linear format, but I think we live multiple lives simultaneously. So when we cease to be physical, all of those lives coalesce together; all of the knowledge comes together at one time. And the reason we assume the light to be God is because all of the knowledge coming together is so overwhelming that we just assume that this must be what God is.

It's the initial threshold of something that we call life after death, but it's the leading edge of the loss of identity. The reason that we return from the near-death experience is as a survival mechanism that says we can't quite lose our identity. True life after death is a loss of identity. It's a reintroduction into whatever the purest form of energy is, that all of creation or matter is made from. Do we become the origin of another star or something like that? I don't know.

My consciousness is scattered across space-time because space-time is an illusion. When I cease to be physical, when I die physically, I cease to be physical in all of those manifestations. All of that experience comes together simultaneously. Now, the reason for being physical is to collect knowledge, or to collect experience. If that's true, you and I are having an experience now. Well, if that's true, then the experience I'm having over here and you're having there is pretty poor because I'm only getting half of it and you're getting the other half. But what if in actuality, we're

both the same? Then we're getting all of it. But we don't
know that until we cease to be physical.

In the physical sense, we don't understand that,
but we have to have the experience by playing
out our roles. In other words, we're incarnated in
multiple lives in the physical. Through the death
process, or the leaving of the physical, all physical
reality ceases to be. All the manifestations cease to be
simultaneously, and it's all brought together into an
understanding of the universe.

McMoneagle's experience is another example of how the trauma of
death experiences can be shifted into a positive encounter that suggests a broader sense of what is possible beyond our embodiment.
And like others who have reported having near-death experiences,
McMoneagle found a doorway into personal growth that leads away
from a primary focus on self-enhancement and toward supportive
and meaningful social behaviors. As research on the positive trajectories of terror management and worldview transformation suggest,
the experiences guided him to reprioritize intrinsic over extrinsic
goals and priorities. It shifted his view from a small "me" to a larger
interconnected "we." As he explained,

It's important to understand that what I do to you,
particularly in this moment, I'm doing to myself.
That's the critical understanding of it. So real karma
is everything you do, you do to yourself. That's the
truest form of understanding. Everything I do to every
living thing, I do to myself.

That's the truest understanding because when
you step out of physical reality, all physical reality
ceases. You're no longer a part of it. You relinquish
your identity. And what's interesting to me is that no
matter what theological or belief structure you come
from, the only real basic argument becomes the basic
argument for identity.

## The Near-Death Case of Pam Reynolds

Perhaps one of the most dramatic cases of a near-death experience is that of Pam Reynolds. She had a brain aneurysm in a portion of the brain that was difficult for surgeons to operate on. The surgeons were concerned that if they tried to operate in that area and failed, the aneurysm could burst. Pam likely would be dead before they'd be able to repair it. So they used a radical surgical method that had not been widely documented at the time. Dean Radin, senior scientist at the Institute of Noetic Sciences, explained the procedure:

> First, you cool the body down. Then you open a vein. You basically take most of the blood out of the body. In particular, you put [the patient] on a table to tip them so there's no blood left in the brain. That reduces the blood pressure in the brain. Then they can do the surgery, repair the aneurysm, and let the blood come back in.
>
> When you do this, and the body is being cooled down to sixty degrees Fahrenheit, it eventually goes into shock and then cardiac arrest. That's like one level of dead. You drain the blood out of the body, then the body is really dead. It's like double dead . . .

"The surgeons then trigger very loud clicks to sound through headphones that the patient is wearing," Radin continued. In a typical patient, the clicks would result in a brain stem response. But if there was no brain stem response, the patient is considered "triple dead."

> There's no blood in the brain. The brain is not electrically active. You're under general anesthetic. You have these loud clicks in your ears. And the eyes are taped shut. It's every form of dead that we know is dead.

In the case of Pam Reynolds, being triple dead triggered a full-blown near-death experience in which her consciousness left her body. Radin continued to describe the remarkable experience:

She saw the tunnel, the meeting of all the relatives, all that stuff. Which is interesting, but still not evidential in the sense that we can't check it against anything.

However, she did say a couple of things that were unusual about the nature of the saw that was used to open her skull—a certain sound that it made—and music that was being played during the surgery, which she reported afterward. It turned out to all be correct. The . . . instrument used to open her skull was a strange shape and didn't seem like it would be the sort of thing used in this surgery. People went back later and thought she must have made it up or she was wrong. But it did turn out that she was right.

The criticism for this as evidence [of an out-of-body experience] is that maybe she actually got some hints about it when she was going under, originally under general anesthetic, or after she was coming back. That's not what her experience was. Her experience was: she woke up when she was dead.

One of the reasons to believe that her experience was unusual, at least, is that when you are going under general anesthetic, or when you are getting hypoxia, or any of the usual physiological shutdowns, your ability to have a clear thought very rapidly diminishes. Your ability to remember what happened also goes away. What she's reporting is completely the opposite of that. It was even more vivid. She remembered the whole sequence [of her surgery] very clearly. So those two aspects don't really jive with the idea of a physiological shutdown that created fantasies [that] happened to match what was really going on. It's one of the more startling cases where the evidence that she was really dead was very good.

## Shifting Motivations for Existence

Dannion Brinkley is an author, speaker, and hospice volunteer in the area of spirituality, self-development, and complementary and alternative medicine. Brinkley's understanding of and compassion for those who are suffering is anchored in his own personal experiences. He's had three of the most complete near-death experiences ever recorded, and endured unimaginable pain. In an interview with me, he explained what happened the first time he died.

> One day in 1975, I was struck by lightning. I was dead for twenty-eight minutes, completely paralyzed for six days, partially paralyzed for seven months. It took me two years to learn to walk and feed myself.
>
> [Just after being struck,] I was in a place that was between this world and the next. But I knew it better than any place that I had ever been living in this physical life. I was comfortable there. I was detached from the event. I was an observer, not a participant.
>
> I watched them load me in the ambulance. I watched the things that went on. I really didn't care because where I was floating above it was so much better than where I was when I was involved in it. And I was in the ambulance, and the guy said, "He's gone. He's gone." Me, I thought, "Gone where? I'm here."

These experiences gave Brinkley a sense of comfort in the face of death. "I grew up in fundamentalism, [which said] everybody goes to hell. . . . There's no question about it," he told me. "But when you realize that you don't die and that you don't go to hell, your whole motivation for existence changes."

## Converging Worldviews

Like Brinkley, Eben Alexander III had a dramatic near-death experience that shifted his worldview. Alexander has been an academic neurosurgeon for twenty-five years, including fifteen years at Brigham and Women's Hospital, Boston Children's Hospital, and Harvard Medical School. Over these years, he has personally dealt with hundreds of patients suffering from severe alterations in their level of consciousness. Many of those patients were rendered comatose by trauma, brain tumors, ruptured aneurysms, infections, or stroke. In his academic career, he authored or coauthored over 150 chapters and papers in peer-reviewed journals and made over 200 presentations at conferences and medical centers around the world.

Given his medical and academic acumen, he thought he had a very good idea of how the brain generates consciousness, mind, and spirit.

> I was kind of back and forth a bit . . . as I spent over twenty years in neurosurgery. . . . I didn't understand how there could be any kind of afterlife because I didn't see how consciousness could survive the death of the brain and body. . . . I believed that I knew how the brain generated consciousness and I didn't see any way that there could be survival of consciousness after bodily death.

Then things shifted suddenly for Alexander:

> At four thirty on the morning of November 10, 2008, I woke up with severe back pain. They rushed me to the hospital, deeply unconscious, and quickly determined that I had bacterial meningitis and about a two percent chance of survival, with no chance of neurologic recovery.
>
> Deep in the middle of coma, my first recollection is basically that I had lost all knowledge of my life before. And somewhere along that way materialized this beautiful spinning light that started out very small and then slowly spun around and came towards me and got more complex, with these lovely tendrils of white and gold. And it was associated with a lovely melody. But as that spinning white light of melody came closer to me, it opened up and became a rip in the fabric, and it was a portal into a whole new realm that was ultrareal, crisp, vibrant, alive, beautiful.

Ever the scientist, Alexander questioned his noetic experience.

> I developed nine different hypotheses to try and explain
> it as a brain-based phenomenon. All of those ideas
> fell woefully short, because I realized that my . . . very
> severely damaged meningitis brain should not have been
> able to have anything like this at all. There was too little
> cortex still working to be able to come up with this
> hyperreality that I witnessed—that very rich and vivid
> reality. . . . I struggled for years with how does experience
> and how does memory happen outside of the brain.

Before his near-death experience, Alexander had hoped there was a loving God and an afterlife, but these questions were overridden by his scientific training. Since his experience, he has been speaking and sharing his spiritual insights with audiences across the world. His story is recounted in his book *Proof of Heaven,* which spent months at the top of the *New York Times* best-seller list. His personal transformation has led him to speak for shifts in society and healthcare, and he has sought to share his message of optimism about death and its contributions to how we live out to the world. While Alexander's experience, like those of others in this chapter, does not offer objective proof of an afterlife, it had a profound impact on his understanding of what it means to be alive.

## BORN TO WITNESS LIFE

As we saw in chapter 1, not all worldview transformation experiences arise out of pain and illness. And not all noetic experiences involve encounters with death. For some people, they may happen in nature and in times of bliss or ecstasy.

Yassir Chadly is an imam (spiritual leader) at the Masjid Al-Iman, a multicultural, Sufi-oriented mosque in Oakland, California. As a young man, he was a swimmer on the Moroccan national team and would often go to the ocean and bodysurf. One day he had a powerful experience out in the sea.

I wanted to surf, and the ocean was like a mat, no waves. So I thought, "What am I going to do? I can't surf. But I had come here, so I might as well swim." I laid on the water on my back and felt those little waves. I relaxed my body in the Atlantic waves, and my body went slightly up, slightly down. My eyes were closed.

Involuntarily, all of a sudden, I felt my body growing out of its limits. I couldn't stop it. It was like yeast rising, and it was growing and growing and growing. I couldn't retrieve it; I couldn't make it small. It was just growing and growing, until me and the ocean were one. I could feel the ocean moving on earth, and me with it. The whole ocean, I was one with it, and I heard inside of my head the verse from the Quran that says, "Say that God is one." And I understood what that meant because I felt that oneness.

I understood that whatever put us here—me, or the ocean, or the sand—they are one. There is no difference. We say "I" or "you" even though we do it just so we can communicate. But in that realm, there is no "I." There is no difference between "I" and "you."

Chadly's experience of oneness and interconnectedness eliminated any divisions in his experience of reality—an experience predicted in the worldview transformation model. While it happened in a moment, the implications have been long-lasting. Over the years, the imam has wondered how his transformative experience relates to what happens after death. He explained:

The more you age, the more spirituality becomes a reality for you. In the beginning, you just do it because everybody else is doing it. As you age, you see it is going to be useful to you for another life hereafter.

As you get older, these questions become more realistic. The more you see gray hair growing, the more you understand. Is death the end of everything, and then there is nothing? Is there something? In the beginning, you see death is coming, but you think it will never happen to you, "Death comes to everybody else except me." As you age, the reality of death becomes real.

I know that that last day is within me. I have that last day, but how am I going to deal with my last day? How is my last day going to be? You start thinking about that more, and then your spirituality becomes more attuned because you start to understand. You are starting to understand pre-destiny. You are surrendering to that, surrendering to God.

There is that fear to go between life and death, about what will it be like. They say that death is everything hurting at once, and this gives you compassion for all mankind. If you see a human being dying, you don't think, "I am Moroccan. This is an American." No, you think, "This is my brother."

I work at the pool, and I open at 5:30 a.m. One morning, I opened the pool and saw that it was not covered. Then I saw that there was a man face down in the pool. And then I saw the gun at the bottom of the pool. He had shot himself. I saw that he was dead, but I didn't think, "He is American. I am Moroccan." Instead I thought, "I am him, and he is me."

Understanding the unity of life led Chadly to ask the big questions: What makes us die, and will there be someone to embrace us after death? For him, the answer comes through the life of the soul.

After we die, the soul asks questions about the work of God. They say that we come from a big funnel—tight at the bottom and open at the top, [with] holes like a beehive. Each hole has our souls in it, so we have a precise place in that funnel. If you are open to the light, you live in an open place in the funnel. If you are not, you live in the dark place. We have a precise place in that place.

## FEELING THE SPIRIT

We need not hold a spiritual worldview to have noetic experiences, nor do our views of the afterlife automatically shift after having such an experience. Michael Shermer offered an example. Born and raised as a fundamentalist Christian, Shermer is now an atheist and the editor of *Skeptic* magazine. He does not require a spiritual framework or noetic experience to cope with painful losses. He shared some of his own worldview with me:

> My father died suddenly from a heart attack. I wasn't there. When my mother died, it took her ten years because of her brain tumors. I was her primary caretaker for that. Then both my stepparents . . . went slowly, so I . . . drove them around to doctors' offices. I went through all that. I'm sure my telomeres are much shorter from the stress of being a caretaker. . . . That brings home the reality of death, for sure. But they're gone; I mean, absolutely I don't think they're anywhere else.

But Shermer surprised me when he acknowledged having a noetic experience after the death of his mother.

> Occasionally afterwards, I thought I heard my mom's voice, that sort of thing. But, you know, we all have these auditory hallucinations a little bit. It went away shortly after that. So it's just part of growing up and the cycle of life. I guess that's how I look at it. I just don't dwell on it.

I asked Shermer how he explained hearing his mother's voice. He replied:

> Oh, I don't want to make a big thing about that. I just remember a couple times in bed, sort of late at night. You know how you're sort of falling asleep? It's just one of those transition things where your mind is producing all sorts of things; I don't think it's anything more than that—anything more than the other hallucinations I've had from sleep deprivation, from sensory deprivation tanks, from nothing more than that.

Perhaps the only real difference between Shermer's experience and those of Jean Watson, Eben Alexander, Pam Reynolds, Dannion Brinkley, or Yassir Chadly is the meaning they ascribe to those experiences. Unlike the others whom we met in this chapter, Shermer dismissed the experience of hearing his mother's voice as a standard hallucination, rather than exploring its significance in his life. His dismissal of meaning regarding his noetic experience does not make him a pessimist, however. He described his approach to life as one of engagement. His role as a caregiver gave him an intimate understanding of life and the chance to serve those he loved. As we learned in the context of terror management theory, his self-esteem and confidence may offer buffers against the terror of death and the attitude Shermer holds toward his own mortality. He illustrates that religion and spiritual practices are not the only ways people feel comfort about their own mortality.

## GLEANINGS

Inviting and integrating noetic experiences into our own worldviews can help us create buffers against the fear of death. This can be true regardless of the evidential nature of nonordinary realms of existence.

How we respond to direct encounters with death and existence beyond the physical is very personal. The people profiled in this chapter offer different ways of knowing and different ways of identifying what is true for each of us. Developing the capacity for equanimity in the face of difficult or out-of-the-ordinary encounters can lead to feelings of connectedness and balance. Questions of life beyond death can be met with curiosity and inquisitiveness, and our questioning may occur within or outside of spiritual practices. Having an experience with death can have profound transformative potential, regardless of our spiritual orientation. Perhaps the great work of the twenty-first century is to develop our capacities to hold multiple ways of knowing and appreciate the diversity of perspectives that make up the whole of our shared human experience. Each of us brings a piece to the puzzle. There is no one experience, no one ontology, that trumps all the others. This pluralistic view may invite each of us to bring a sense of humility, compassion, appreciation, gratefulness, and interconnectedness to our relationship with death and what may lie beyond the physical. It is this range of perspectives that will be explored more fully in the following chapter.

◄ PRACTICE ►

## Connecting to the World

Find a comfortable and safe place to sit and relax. Take three deep, slow breaths and feel fully present in this moment. Feel the chair you're sitting in, feel the clothes on your body, feel your breath going in and out in your nostrils.

Now close your eyes and imagine yourself in your favorite location in nature. It could be on the beach, on a hiking trail, in the woods, or next to a river. Feel fully present in that place. Feel your toes in the sand or water or grass or dirt. See what is all around you. Imagine the air on your skin: Is it cold or warm? What are the sounds you hear? What smells are in the air? Experience the joy you feel as you experience nature in silence for a few moments.

Still imagining yourself in your favorite location, feel the connections between your body and the world around you. From this place of interconnectedness, consider your worldview on death. Have you had any noetic experiences? How did they inform your own perspective on death and the afterlife? What do you think happens to your personal identity at the point of death?

When you are ready, return your mind to your room. Record your observations in your journal. Reflect on the transformative nature of placing your attention on your interconnectedness with nature and expanding your thoughts beyond your body.

# Chapter 4

# COSMOLOGIES OF LIFE, DEATH, AND BEYOND

Whether you're Christian or Buddhist
or Muslim or atheist, everyone meets
at that point we call death.
YASSIR CHADLY

At the time I met him, Gilbert Walking Bull was an elder of his Lakota tribe. He saw himself as a bridge-maker between worlds. He grew up in a small village on the Pine Ridge Reservation, at the edge of the Badlands in South Dakota. Raised by traditional holy people, he was chosen by the spirits to serve Tunkasila, or the Great Spirit. Walking Bull was trained by his grandfather, who also worked in the spirit world (*wanaoi,* in the Lakota language). His grandfather, like Walking Bull, "was color blind; he healed all people, and people respected him."

It took courage for Walking Bull to share his worldview with those outside his tribe. Many of his people would not have wanted him to communicate his cultural knowledge to the non-Indian world. Yet he felt that the Great Spirit guided him to impart the healing power of his tradition "so it may help others down the line."

Walking Bull described his unique childhood and being raised off the reservation.

> It is where all the holy men lived at one time. They
> worked together. They didn't want to lose our Lakota

culture and religion, so they found kids that they kept out of the reservation schools, and they raised them in the traditional Lakota ways. I happened to be one of them. I never went to school, but I have been in ceremonies and guided by my grandfathers. . . .

In order to understand these things, you must study to become familiar with and reach the unseen world. To reach the unseen world, you must come from your faith and belief within—not from a book. But to learn from people who have a strong religious or spiritual path to teach and guide you how to concentrate, pray, sacrifice, then you will feel the Great Spirit's energy and power.

It is hard for me to explain the name of my religious path. The being who created the whole universe, whose name is Tunkasila, is Grandfather. When we address the Supreme Power, we call him Wakan Tanka. Great writers didn't and don't understand what *Wakan Tanka* means. The holy man was never asked to explain what these words mean; all they knew is it meant to meet God, the supreme God.

Wakan Tanka is held by the Lakota as the direct connection to the creator of the universe. Walking Bull continued:

*Wakan* is your artery that connects your heart to your mind. In the same manner, it is an artery that connects our mind to Great Spirit. The energy of that power is in the center of our brain and is activated through how we transmit our prayers through concentration. We concentrate our thoughts, our prayers, and they travel through the center of our head to Grandfather. So that is *Ka,* the unseen power through which we are connected. How you view yourself and concentrate to focus on these powers is often a result of how we were raised to understand ourselves clearly. I grew

up knowing these things. Sometimes I would pray
for someone I felt sorry for. I'd concentrate so that
my spirit would travel to the spirit world where it
communicates with Great Spirit to heal that person.
So I always had the feeling inside, which is how
I can explain to others how to pray to the Great
Spirit—connect to God—to the power that exists.

The Native American elder also explained his worldview about death,
noting his own caution when it comes to dealing with the spirit world.

Any person who wishes to know ghosts or common
spirits needs to know that they do exist. There are
good ghosts that exist and bad ghosts that exist. We
know this, and so we protect ourselves from this
world. When ghosts contact us, it is very dangerous
for us. I am sure other people believe they see their
dead relations and other type of spirits. I respect
them [spirits], and I don't try to deal with them, but
I pray for common spirits. The Great Spirit has a way
to control spirits on the other side. He has an unseen
power—a dimension—that blocks the common
spirits, so they can't come through; they can't talk to
us. But if you are a very sensitive person, you may
see ghosts, common spirits, out of the corner of your
eyes. They won't come in front of you where you
focus your power. You project power through your
eyes. The common spirits respect that and will not
show right in front of you. So only the sacred spirits
will appear in front of you.

## A MEETING OF WORLDVIEWS

Meeting with Walking Bull in the early days of the twenty-first
century brought home to me the uniqueness of this moment in history. Never before have so many worldviews, belief systems, ways of

engaging reality, come into contact. We have all the enormous successes of science and technology. We celebrate the ways in which we have been able to master the physical world—cloning cats, taking personal computing into our pockets through smartphones, and manning an international space station. We also now have access to the world's wisdom and spiritual traditions through a few strokes of a computer keyboard or a flight across continents. The diverse ways that humans can express themselves are now coming into contact at an unprecedented rate.

Such are our lives. On the one hand, studies in the terror management theory literature, such as those reviewed in chapter 2, reveal the ways in which mortality prompts can lead to strong in-group identification. This identification, in turn, can lead people to oppose others who view the world from an alternative worldview—in short, conflict. And religious and ideological wars have been responsible for massive deaths and destruction. On the other hand, people believe many different things about reality, and by and large, we get along. As we hold different worldviews, we have crafted ways to coexist. Christians, Jews, Buddhists, Hindus, Muslims, and atheists are all sharing the same grocery stores, medical centers, and schools. And at the same time, they adhere to different ways of understanding our human experience. As we have also seen, raising death awareness in the context of compassion, curiosity, and cultural appreciation can actually reduce defensiveness and antagonism toward others.

The lenses of diverse spiritual traditions provide us with resources for transforming the fear of death. Rupert Sheldrake is a biologist and advocate for a new, more spiritually evolved science. He shared with me his views on death:

> I think contemplating dying is essential to how we live.
> If you're afraid of death or you're in denial of death,
> then your life has to be a kind of evasion of death, or
> keeping yourself so busy you can't think about it.
>      I think one of the great advantages of most
> traditional religions is that they make people less
> afraid of death. Death is scary, especially if it's painful.

But they make people less afraid of death because there's a feeling that death is a transition. If you believe that death is the end and the mind just goes blank, you may be scared of that or scared of growing old. So anything that makes people less afraid of dying is probably a great help in the way that people lead their lives.

In truth, it is hard to find anyone who is exactly from one tradition. Cultures blend. They inform one another. Each of the wisdom holders that I spoke with is influenced by his or her own and other traditions and customs. They are bridge people who walk between worldviews, tending their own while holding a sense of curiosity and appreciation for alternative perspectives.

What follows are some diverse perspectives from a range of cultural worldviews that illustrate various beliefs on the questions of death and the afterlife. We will again hear from Yassir Chadly, a Moroccan imam who plays rock music. Lewis Rambo is an evangelical Christian who has moved beyond his fundamentalist upbringing to study Islam. Rabbi Jonathan Omer-Man is a Jew who celebrates Buddhism. Qigong master Mingtong Gu is deeply drawn to genetics, and Rick Hanson is a neuroscientist who practices meditation. The twenty-first-century worldview is clearly a bricolage of the most compelling and powerful type. By embracing the possibilities within this dynamic pluralism, we may gain potent new insights into our relationship to life, death, and what we believe comes after.

## Nature and Spirit

There are many traditions that base their spirituality and worldview about death in a naturalistic context. In my efforts to connect to this embedded view of humanity and nature, I spent time in the Ecuadorian Amazon. Deep in the forest, I was guided by Santiago Kawarim, then president of the Achuar Federation. Wearing bright red face paint in stripes across his rounded cheeks and a brightly colored feathered headdress, he spoke to me (through a translator) of his passion to preserve his indigenous people, the Achuar, while moving them forward into modernity. Maintaining a relationship with those who have passed over is an important part of his worldview.

> In our world, we believe that when a person dies, they become another animal. When someone has an experience with a spiritual force, he is transformed in order to transmit the force to another person, in order to help them. But generally speaking, we believe that when a person dies, they transform into animals such as the owl, the deer, or the pacifist bird. Also, these animals can transform another person and take them [into the spirit realm].
>
> For example, if a person has been transformed into a deer or a pacifist, they can enter into or be able to see how things truly exist in reality. They could see their dead grandparents, grandma or grandpa, or dead brothers and sisters. Such cases have happened among our people, even to children. For example, the deer has taken children to their elders who have passed away. At times, a deer appears and comes closer, without being scared. And then the deer takes over the sphere of the person who is still alive. Once the energy of the deer envelops you, you are able to see a different world. And once a person has entered into that other world, he's able to see this deer as a person. So a person who has not been transformed to see the deer, sees only a deer. But when a person is close to the deer, then they see a person, like a dead grandparent.

The Achuar believe that the soul leaves the body during sleep or when a person is using the medicine of a plant like *ayahuasca*. In these extended states of consciousness, the soul interacts with the spirits of the forest and spirits of departed loved ones. The Achuar gather in small family units in the early morning hours to share the content of their dreams, exchanging information that they gained from the spirit world in order to guide their daily actions. By engaging the spirit realm, they gain important insights about life. Their dream life is rich with symbolic messages from nonphysical dimensions. With the increasing impact of the modern world on this previously isolated society, however, many Achuar are losing their close connection to the spirit realm. News programs are broadcast on the radio at the time when people previously shared dreams, creating a new way of connecting to the world beyond the rainforest, but pulling the native people away from their connection to the spirit world.

## The Hoop Dance

Many indigenous people believe we live in a circle that revolves around the four directions. To better understand this worldview, I sought out Tony Redhouse again. He is eclectic in his worldview, and the power of his message has supported people in hospice work, those in substance-abuse recovery, and those suffering from cancer and emotional trauma. He described his understanding of life and death:

> In the Native American tradition, we believe the circle of life revolves around and around. There are four different points, four different winds, four different directions. . . .
>
> When our mind and body are strong and healthy, when our soul and our prayer life are vibrant and alive, when we have a connection with the spirit, and when all of our relationships are at peace, that makes a nice, balanced circle. When the human life cycle connects with the animal life cycle, and when that connects with the plant life cycle, and when that connects with the lunar cycle, what happens is that you begin to make designs that represent what would be the universe or the world, which is one big circle. And all of these circles are making the circle of life, the world, the universe, complete and whole and a place of harmony. . . .

The hoop dance is basically connecting all circles
of life together to create bigger designs that are
representative of my journey in this life, my connection
to the eagle, to the hummingbird—the wisdom that I
have learned from these different symbols, these different
animals, life forms, that I can apply to my own self. So
all of these designs that I am creating are kind of an
unfolding and a picture of my life journey and how I've
connected with all different life forms to gain wisdom.

## Death as a New Beginning

Like Redhouse, Luisah Teish honors the cyclical nature of life and
death. Raised in New Orleans, Teish has been steeped in practices
that blend many traditions from the African diaspora. A practitioner of the Yoruba tradition, she holds the rare title of *yanifa*. This
word means "a mother of destiny." It implies that she has progressed
through several stages, gone through a number of initiations, and
learned enough rituals that she is considered wise enough to guide
others to fulfill their contract with creation. She explained her tradition's view of the dead:

> You see, the dead are not under the earth. They are
> in the water. They are in the woods. They are in the
> fire. They are in the air. They are in the breast of the
> woman. They are in the child that is crying. The dead
> are not dead.
>
>   For folks in my tradition, we believe that you can leave
> your body and come back into it, in this incarnation.
> What we call *death* is simply leaving the body and
> returning to the land of the ancestors. Most often, there
> is a commitment to come again. We're not really trying to
> avoid earth life. I know some people think of life on earth
> as some sort of a curse that you want to avoid. No.

In Teish's animistic worldview, each of us is connected to spirit that
always has been and always will be.

In my belief system, before I came here, I was in the land of the ancestors, and when I leave this body, I will return to the land of the ancestors. The whole time that I am here I am continually in contact with the ancestors and what's going on. That is where I get a lot of my assistance.

We say the world is a marketplace and heaven is my home. So I go home and rest, and I come back into the marketplace. And then I go home and rest, and I come back into the marketplace. That connects to the idea that every person who actually makes it through the birth canal, and who chooses to stay, came to be a particular person, to do particular things, and to have certain experiences.

We come in having a contract with creation, and fulfilling that contract is what life is all about. I'm the kind of person that people come to for guidance around that: "How am I doing with my contract?" "Did I get off the road?" "How do I remove this obstacle?" We ask for a death that is appropriate to the archetype that the person is living under. It is our belief that every human being is a child of a particular force in nature.

Teish is an intermediary between the living and the dead. She serves as an *ip'ori,* or person who helps connect people's earthly heads with their heavenly heads.

An ip'ori knows what your contract with creation was when you chose to take a body and come into this world. In the process of being born and being socialized, we do the best that we can to remember what we can of the original contract, but we can get misdirected or redirected through chance and choice.

## Reflecting on Impermanence: Embracing the Qi

Mingtong Gu is a qigong master and spiritual teacher. He is a man who bridges his traditional upbringing in mainland China with his busy life in modern-day America. He shared with me his ideas about the diversity of views around death:

> I think there might be some generic difference in the attitudes and understanding of death in the East and the West, and yet each tradition is so different. I won't generalize East and West. There are so many different traditions in the East, so many different traditions in the West. There may be more about the ancient connection compared to a culture with a contemporary way of looking at life, more distant from the ancient link. On that level, I feel some difference.
>
> In the ancient culture, there's more honoring death, more embracing of death. There are different kind of rituals. People go slow when they lose loved ones. For example, in China, people have slow grief for forty-nine days. Every seven days equals a ritual, not only for the living in the releasing of grief, which can be hard, but also as another way to honor the people who passed on—for them to dissolve spiritually. That is a very wonderful ritual.
>
> In the more contemporary culture, we have a hard time dealing with death. In general, death is the most frightening thing to deal with, so different cultures developed different rituals to face that ultimate challenge. Finding the peace in death, finding the meaning in death, is very important for any culture, for any human being.
>
> From that deeper level, the ritual for death is as important as the ritual for birth. And without death there is no birth. Without death there's no life. Scientifically, we know there's incredible life and death happening every moment in the cellular level. Different cells go through different cycles through death and life, regenerating again. Death and regeneration.

People say that in seven years the whole body goes one full cycle. Seven days another cycle. Seven days, another cycle. And deeply, instantaneously in the quantum field, the DNA is vibrating in these cycles. Winding and unwinding again. That is ultimately the deeper level of death and life happening in the body. I think we here are continuously learning, creating a certain vehicle, not only for people who pass on, but also for people living to deal with death as a very meaningful lesson for life.

## Changing Clothes

For Satish Kumar, whom we met in chapter 2, death is not the end of life, but a new beginning. The social philosopher and ecologist used practices and metaphors to express his view of death as transformation.

When your old clothes get worn out, are you afraid of changing to new clothes? In the same way, we should think that when our body is worn out, we should not be afraid of getting into a new body. So, from the Indian perspective, life continues. Life and spirit never end; we just change bodies.

Death will liberate you from this old thinking, old mind, old relationships, and old habits. It's a liberating force. We should look at death as a liberation. If somebody says to me, "Satish, I give you boon—you will live forever," I would say, "I don't want to live forever. I want to live a new life!" So for me, there's no fear of death.

When the autumn comes, the leaves are brown, and then they fall. Where do they go? They go back into the soil. And what happens to those leaves? They biodegrade, and they become soil. Then the nutrition of those leaves, which become like compost, goes into the roots, and comes back into the tree and the trunk. Then here comes the spring, and the new leaves are born again. So if the old leaves are afraid of dying, how will the new leaves come? People who don't believe in reincarnation have to think that life continues. Life never ends. It's only the form that changes.

For Kumar, being intimate with living and dying brings us directly into a transformational process that follows an organic cycle—the cycle of life. Each of us engages that cycle at a different point and in our own way.

## The Song of Islam: Death Is Life

As a Sufi imam who embraces pluralism, Yassir Chadly finds the topic of death to be unifying and universal.

Whether you're Christian or Buddhist or Muslim or atheist, everyone meets at that point we call death. That's in general. But in the spiritual world, in the Quran for example, it is mentioned that the Prophet, peace be upon him, said, "Blessed the one who created death and life." Usually you think life and death; we call this life and [think] that death is the end. But in that teaching, he calls this whole life death, and he calls the after-death life. Because this one has a limit. You are born. You have an ending. The other one doesn't have an ending. And he calls that "life," and he calls this place "ruled by death." That's what they call it.

And more than that, he said he created it: "It's not your business, this death and life. It's not in your hand. You pass through it because they are creation, as any creation." He created the elephant and giraffe and zebras, and he created human beings. He created death and life. So he said, "Don't mingle with this. It's my creation, I created that." So what's left is how we are going to respond to that.

I asked him how we can respond to death. Chadly's views offer a blend of rigorous Islamic practice with the fluid Taoist concept of yin and yang.

> If you understand [death] like that, then the response is really positive because you are now going to a place where you live forever. One of the beautiful words you have in English is *eternity*. What is that word, *eternity?* We say this life doesn't support eternity. Therefore, there must be a life that is eternal because of the yin and yang and the balance. If one doesn't have it, the other one has. So they balance each other. The reason we have what we call life, and Allah calls the death, is to test you. How are you going to respond? And so we're here to be tested and choose. What choice you are going to take here is what is going to reflect in the hereafter, in the eternity place.

As a devout Muslim speaking from an Islamic worldview, Chadly is part of a demographic that numbers approximately 1.6 billion people in the world. Having said that, each person represents a different understanding on these complex ideas. There is no one worldview that defines all Muslims. Chadly explained that his cosmology, based on Islamic teachings, includes both predestination and the choice of good or evil. Chadly described the latter this way:

> Sometimes you go to a bakery. They have little sample tastes. So this whole planet is a little atom, and we are in it. The planet itself is a sample. And in it, everything you do is a sample. So if you do bad things, and you taste it, and you say, "I want to do more of that," then in eternity Allah is saying, "I have a lot of that if you like." And that's what we call hell, because you wanted that.
>
> So it gives you more of it there. And if you sample good things, and you say, "I choose this,"

he says, "I have a lot of that, too." So he gives
you a sample of both of them and tries you. He
gives you a little bit of bad things and little bit
of good things and says, "I want you to know
which one you like. I know what you're going to
choose, but I want you to know yourself." And
so, people sample; and in the hereafter, they find
what they sample over there, a lot of it. And
that's the test.

"What happens when we die?" I asked Chadly. When the physical
being stops, what would he say is next? Like Satish Kumar, who was
raised within the Jain tradition, Chadly used the metaphor of clothing to describe our embodiment:

Our physical being is like a dress. We just wear this
physical being. Our soul lives in this physical being.
When you leave, you leave your clothes. It's like
when you swim in a lake, and you put your clothes
on the shore. As you swim . . . you look to make sure
nobody's going to steal your clothes. So you have a
connection with your clothes. Our souls also have a
connection with the body. Even when we leave we
can say, "I used to live there." We used to live in that
place. That's how the body's going to look from that
side, [to] the soul . . . After forty days are passed, the
angels bring one ray from the heavens, from your
soul, and they put that in that place where the souls
are gathering.

## Exploring the Mystical Side of Judaism

Rabbi Jonathan Omer-Man is a religious scholar who finds common
ground with Christian monastics and is drawn to the study of Islam.
He moves in a wheelchair, and he is finding himself winding down
his career. While he has no intention of abandoning his Jewish
path, he holds a worldview that appreciates the multiplicity of faith

perspectives. When asked about the goal of his own practice, he was able to take a wide lens:

> I don't like results or benefits in spirituality. For myself, it's the ability to move from the exoteric to the esoteric. It's the ability to reach the place of unity of all spiritual quests without obfuscating, without ignoring the complete separateness of the different paths. There is a transcendental unity of religions. The ways of getting there are quite different. There is no way in which you can combine a Christian Eucharist with a Jewish Torah study. Possibly the level of being or the knowledge of the divine that comes at the end is the same. There is a level at which one moves with ease and the knowledge of all, the knowledge of the totality, and one doesn't forget the road where you got there.

Within the Jewish tradition, human lives will be measured after death by whether they have lived up to their fullest potential. The time of aging and death awareness involves reflection, of reviewing the deeds of service in one's life. The concept of a world-to-come, or life after life, referred to as *olam ha-ba,* is central to Judaism. This is a time when people will be judged for their good works. Death is not a punishment, but a natural part of the life cycle. When a person dies, their soul lives on. It has both awareness and consciousness, according to the central teachings. Life in this world is fundamental to what will happen after. Reflecting on his changing life circumstances, Omer-Man shared his own reflections on whether his was a life well lived.

> A few years ago, I came to the realization that I wasn't going to finish this time around. [There is] much too much left to do. I feel in recent years that much of my work is cleaning up the mess of my earlier years. [I am] trying to leave this state slightly better

than when I came in. In some ways it is much less grandiose and much less ambitious. I think I've become lighter. I think I've become less stern and intimidating. I think I laugh more. I feel much more of lightness of being, and of course I have a lot of work to do without mobility, looking at the transition to the next stage. In our society, you're not meant to do that. You don't think about your own dying; and if you do too much, you should take Prozac. I have become a more reflective human being.

## The Wide Spectrum of Christianity: Beyond Heaven and Hell

Within American culture and throughout the world, Christianity is an important religion. There are about two billion Christians in the world, making up about 33 percent of the global population. To understand better how some people hold death within the tradition of my own childhood, I found my way to the doorways, pews, and altars of teachers from diverse traditions within the Christian faith.

### Questioning Assumptions

Lewis Rambo is an expert on religious conversion. He grew up in the Church of Christ, a conservative branch of Christianity that holds a literal interpretation of the Bible. Today, he describes his relationship to this belief system as ambivalent and complicated. Still, it rooted in him a passion for religious and spiritual practices. When asked to share his insights about the Christian worldview on death and the afterlife, he refrained from making generalizations.

> The standard orthodox answer—and I don't mean *orthodox* as in Greek, but as the mainline view—would be that your body dies and your soul or spirit is an entity that somehow transcends death.
>
> In traditional Catholic terms it's called *limbo*. Or [it's] some sort of state like in the Second Coming of Christ. There will be a resurrection of everyone. There

is a judgment when you die. Then you are sent either up or down (metaphorically speaking) to heaven or to hell. Hell is variously conceived as a place of eternal torment, fire, and all kinds of punishment. On the other hand, if you are on the right side, you go to heaven. There, you will experience eternal bliss. It's visualized in different kinds of ways. In the Book of Revelation, it's the New Jerusalem. The light of God is eternal, and there are angels and saints and everyone is singing the praise of God. That is a good place to go.

There was a lot discussion of heaven and hell when I was young—probably more comments about hell than heaven. . . . Not too many people in those days were drawn to the good stuff; rather, they ran from the bad stuff. Even among evangelicals, there is a belief in heaven and hell, but it's not talked about as much these days. I would say there is a great deal of fear for some people about going to hell.

As a professor of comparative religions, Rambo spoke with humility and candor about his ambivalence about the afterlife and the values that inform his worldview.

What do I believe? I don't really know. There are those times when I have that moment of terror: "What if the fundamentalists are right, and I'm going to hell?" And then I jokingly say, "Well, that's where all the interesting people are going to be. The boring people will be in heaven." Others, like Freud, Marx, Che Guevara and the Dalai Lama, won't be there because they're not Christian. So a lot of other interesting people won't be there as well. That's sort of evading the issue. The truth is I don't know, and this is a real struggle about the human imagination.

I guess if pushed to the wall, and in my best moments, I would say that the creator of the universe is benevolent and that surely a benevolent god would not punish evil people whose evil is, after all, finite infinitely. I remember as a child, this was kind of spooky to think about. I remember going to sleep at night and praying and thinking, "It's going to be forever and forever." And this could go on for thirty minutes, [thinking about] forever and ever—a real profound, deep, earth-shattering kind of terror of what eternity would mean.

If there is life after death, I would be happy if Hitler were simply exterminated, but I'm not sure that I would want Hitler to suffer eternally. Maybe give him ten thousand years for each person killed, but forever? This is just a struggle I have; can a creator god, if that god exists, punish one of the supreme creations that god has created, forever and ever? And if that's true—and this is on the verge of blasphemy—I don't want to be around that kind of god.

From my background, it's scary to even say that because maybe evil is so evil that it has to be eternal punishment. It's hard to imagine that with my feeble little brain and mind and heart. At my best, I wouldn't want even the worst person on the earth to suffer that much. Surely, if there is a God, that god wouldn't be the agent of torture forever.

## The Kingdom on Earth

Unlike Lewis Rambo, Lauren Artress was raised in the progressive end of the Christian tradition. Like Rambo, she was quick to acknowledge the diverse range of perspectives about life and death within the Christian faith.

"The major tenet," she explained, "is that there's a life after death." A central issue of debate in Christianity involves the concept of resurrection. One approach involves a physical resurrection, where you take your body with you when you die. Artress does not embrace this view. On the other hand, she does believe that there's consciousness beyond death and beyond the body. This shapes her view of life and the afterlife.

> There is that sense of something that happens after death. I think that one of the wonderful gifts that Christianity brings to the world is that death is not the end. I think that's what we mean by soul. Soul is the eternal part that lives on after death. And it's truly not only a mystery, but I think it brings a great deal of realistic comfort. It's not sugarcoated. It's something about you; your essence lives on.

In Christianity, as in many traditions, the concept of the soul is important. For Artress, there is a possibility that we can grow our soul in the course of our life's journey.

> I think that's part of the purpose of this life in the body. As we make choices on the behalf of our greater good, find our purpose in life, really come into blossom in our own, find our work in the world, find our gifts . . . to bring to this life. It's always a learning process. There is a spiritual path to it.

We have an opportunity in this embodied experience of life, according to Artress, to inquire into ourselves. She identifies four questions that we may ask as we seek to grow our soul.

> Are you deepening in compassion? Are you learning to be less judgmental? Are you learning to be more patient? And is there a way you are finding to be of service in the world? That's the essence of spirituality. Growing the soul has a whole lot to do with how we answer.
> I think that the soul really can awaken. We know that it's often asleep for many people, and it can awaken. As we become more conscious of this part

of ourselves that we only get little glimpses of, that is eternal, we want to guard that, protect that, grow that, be in love with it.

As we talked together at Grace Cathedral in San Francisco, I asked Artress if she believes in heaven. She shared her view:

> Probably not in the medieval sense that there's a place up there that good people go, and therefore there's a bad place down there that bad people go to. I do believe in heaven, though I think we can have heaven on earth. I think by coming to peace with ourselves, coming to peace with one another, deepening our compassion, lessening our judgments, encouraging our patience, then heaven is on earth.
>
> When we talk about Judgment Day, I think one of the important teachings of the Christian tradition is that judgment is done in love. People have this feeling that, "Oh my God, Judgment Day is going to be a terrible moment in my life or in my death." A lot of people feel that Judgment Day means punishment rather than judged in love. A lot of people feel that they're inherently bad. This concept of original sin either has to be entirely jettisoned or explained clearly. But that sense of being innately bad, I think, is such a condemnation of the human spirit. And that needs to be brought to light because a lot of people sign up for that. And instead we're judged in love, which means we're all released into love, into light.
>
> We're all learning. I think there is such a thing as an evil person. Such a person really cannot understand the gift that life gives us. But that's a very, very small percentage—one percent even. Most of us are kind of fumbling, trying to learn our way, trying to learn what this is all about, trying to discover what our purpose is in life. What does it mean to love? What does it mean to

walk a spiritual journey? As we learn that, I think that it does have a lot of impact on what happens when we die.

## Unfolding Life

As we learned in chapter 2, Michael Bernard Beckwith is the founder of the Agape International Spiritual Center, headquartered in Los Angeles. His teachings embrace the New Thought–Ageless Wisdom of spirituality as taught in the East and West. According to Beckwith,

> in everyday parlance we use terms like *past lives* and *many lifetimes,* and while there's nothing inaccurate with that language, I prefer saying that we have one life with different chapters, which include adventures beyond the three-dimensional realm wherein we're exploring new terrain as well the landscape within our own expanded awareness.
>
> Our individual life is just one beautiful unfolding with an infinite amount of chapters created for the purpose of revealing the fundamental order of the cosmos—the beauty, the intelligence, the love, the joy—according to our unique, individual pattern of unfoldment. This approach allows our life to be more of an adventure, a discovery, rather than railing against or trying to outpace physical death.
>
> Individuals spend so much time trying to prevent death rather than seeking to live life. I think most people are not afraid of dying; they're actually more afraid of truly living.

For Beckwith, death is associated with awakening to the essence of our core being, which includes different types of awakenings.

> There is a type of awakening that takes place when a person wakes up from a long night's sleep and they cognize that they were in dreamland.

Then there's an awakening that takes place when a person leaves their body: they "die," yet they realize they are still alive. They experience how every thought, intention, action, and nonaction experienced in the physical body has now become part of what could be described as their newly acquired subtle body.

When a person dies, it doesn't mean they become enlightened; they simply awaken to the fact that they are no longer in the physical body. They realize how they spent so much time worrying about a death that never happened, because death doesn't exist. And then there comes another dimension of awakening that occurs when a person consciously realizes their oneness with Spirit and all existence.

Accepting the truth about death allows us to fully live our current life with a newly found purpose because we know ourselves beyond identification with the ego.

I believe that upon arriving on the other side, we have an opportunity to review our life as we lived it on earth. This review includes how we honored the karmic and dharmic purposes for which we took a human incarnation. Our review process occurs completely free of judgment and enables us to continue our evolutionary progress in a new dimension of living that supports the ongoing unfoldment of our spirit-soul.

### Matter Can Never Be Destroyed

A former professor of religion at the University of California, Berkeley, Huston Smith has written and taught extensively about the diversity of spiritual worldviews. A Christian, he explained his worldview, which links consciousness, science, and spirituality:

A good lesson we can learn is that matter, which science considers the most fundamental element, can never be destroyed. It can change from corporality to energy—back and forth—but it's impossible to destroy it.

If consciousness is the fundamental element, it cannot be destroyed either. That means that when you drop your body, your consciousness is going to remain. The light on the television of your mind never goes out. Of course, it's impossible to know what images will be on that screen after we drop the body, but the light is going to be on.

Now, somebody may not take this as good news if they've had an unhappy life. What's the virtue of prolonging it forever? But that's the minority, and transformation can change that.

What happens to the ego when we drop the body? All of the religions say that it contains impurities and cannot enter the infinite perfection of the ultimate reality in life. Therefore, there must be a cleansing. In symbolism, this is hell. Let's just take it as I do—that it's a metaphor for cleansing. Fire cleanses, and therefore, the first experience after we drop the body may not be pleasant. As the Muslims say, "Remember the Day of Judgment." We have to face that day, and we have to be cleansed of the soiled aspect. But after that is burned away, then the dewdrops slip into the shining sea, and the ego either vanishes [and] evaporates, or it expands to include infinity.

Personality in that beginning period does not survive after death. The ultimate personality is the infinite. There is a saying that when people are drowning, their whole life passes by them. Fast-forward, I received a letter, and it's this person's experience that came first. Then rewind to the beginning, and this time he went through his life conscious and feeling the pain that he had inflicted on others, and he was feeling as if it was his own pain. However, rewind to the beginning a third time, and this third time it was as if God was saying, "Okay, you did hurt other people a great deal; however, you're only human, and you're forgiven." So those three phases die.

## Being Agnostic

Sam Kean, whom we met in chapter 2, has wrestled with death many times in his life. Like Lewis Rambo, he grew up in a fundamentalist Christian tradition. The death of his father called him to question his own worldview.

[In the religious tradition of my youth,] death wasn't a natural event. Whatever came afterwards was determined by your beliefs and whether or not you believed in Jesus.

I was brought up with that verse, John 3:16: "For God so loved the world he gave his only begotten son that whosoever believe in him should not perish but should have everlasting life." Well, to a kid that isn't very symbolic. But the kicker is, "whosoever shall believe in him." What if I couldn't believe? I could never believe sincerely enough. As soon as you examine your belief, you're bound to find out, by the end of the fact that you're examining it, that it isn't sufficient.

Death had a lot of sting in it for me, and my father's death had marked me quite severely. So I was one of those who grew up with awareness of death, not one who denied it. I was critical of a lot of it all along, but I think that the failure of fundamentalist Christianity to save me from death struck home when my dad died. If it's going take the father, it's going to take the son. So that was the beginning of my transformation in that way. And I went into psychotherapy, and once I was in psychotherapy, the die was cast. I had to systematically destroy my character armor.

I have two very definite and ineradicable and opposite feelings about death. One of them is [that] the slate is wiped clean. From nada into nada. Sometimes we have to remember that the nothing out of which God created the world is the nothing I'm going into in death. And if it's a nothing that created all that, like the Buddhists say, "the fertile void," then you shouldn't worry about it too much.

So, one is that I'm annihilated and the other that I'm not annihilated. I don't believe in a god that doesn't recycle somehow. This is an ecologically inferior god. And I'm not making a dogma. I'm just saying that those are both feelings. And they are the opposite feelings.

When my dad died, I was very much in agony about the whole thing. And one day on my way to work, I just heard this voice, and it said, "You don't have to know." And it was, like, "I don't have to know." It was such an enormous relief.

## Honoring Atheism

As Kean implies, agnosticism is not the same as atheism. The former involves a questioning and a comfort with not knowing the fundamental answers about death and the afterlife. Atheism is a lack of religion that holds the position that there is no God and no afterlife. For many people, it is a way to make sense of their lives. In their worldview, atheists often articulate a humanistic or naturalistic perspective on life and death.

Michael Shermer, whom we met in the previous chapter, is one of the leading spokespersons for an atheistic worldview. He takes delight in earthly activities grounded in his human embodiment. He shared his own perspective regarding death:

It's a bit of a paradox that we can observe it happening around us everywhere. About a hundred billion people have lived before us, and there's about seven billion living now. Every one of them before has died and not come back, as far as we can tell. We can see that, and it seems pretty grim.

We can't imagine that it will actually happen to us for a very simple cognitive reason: picture yourself dead. There's research that's been done on this. And what subjects always say is, "Yeah, I can see it. I can see my body, and there's my casket, and there's my friends and family around it." No. You're still in the picture. You're still observing it. You don't get to do that when you're dead. There's no observer. In other words, you can't really conceive of it any more than you can imagine there is no universe. You just can't. You hit an epistemological wall that brains and sense can't deal with.

So we're faced with that sort of cold hard reality. I think that produces anxiety in some people. It doesn't bother me; I don't worry about it. But I know a lot of people do. Maybe it's just my temperament. Like when Socrates noted that we have to be alive to experience life, and when you're dead, you experience nothing, so there's really nothing to worry about. There is nothing even to think about. I don't worry about it. Just eat, drink, and be merry, for tomorrow we die. No, just kidding. But I'm not terribly concerned about it. I don't think about it a lot.

One of the eye-openers for me when I was a Christian and a believer was the study of world religions and comparative mythology. I love all that stuff. But it was an eye-opener: So good people think just like me, except that their ideas are very different. But they're just as good as I am. Though in a vast sea of human religions, what are the chances that I'm right? Very slim. Probably none of us are right. It's

probably a byproduct of some other cognitive process;
that's what I think.

The idea that there is an invisible agent, I call that
God, and an invisible place . . . we're naturally born with
this [idea]. It just comes naturally to protect ourselves.
. . . From the theory of mind . . . it's just a tiny step to
imagine, because we can't really grasp ourselves dead.
You can't really do it. So it's just tiny little steps to
imagine yourself continuing on somewhere else.

Many people associate religion with morals and values. Anticipating
an afterlife guides our everyday actions. For Shermer, there is no
connection between a person's religious belief system and the nature
of their values or behaviors toward others.

First of all, there's no evidence that atheists are any
less moral than Christians or any believers. So it's
kind of a moot point: Who's the more moral? No, but
why do they act morally? Because, I argue, that we
actually evolved moral emotions that were there long
before religion ever evolved. Hundreds of thousands,
or millions of years ago, as a social primate species, we
evolved these emotions to help us get along because
we had to. So what religion did when it came on is it
started identifying certain characteristics of human
nature that needed to be kept in check—the free-
rider problem, infidelities, truth-telling, these sorts of
things. Religion came after the fact.

Atheists are moral because that's what we do. We
are moral creatures most of the time, and in most
circumstances, most of us are good and do the right thing.

## From Diversity to Pluralism

Integrating all the diverse worldviews about death and what may
lie after is mind-expanding and sometimes confusing. How do we
make sense of so many truth systems? Is it possible to live into a

new worldview that includes the multiple ontologies as they inform some unified whole of our collective humanity? Can raising our death awareness around diverse models of reality help to alleviate the terror of death and ameliorate aggression toward others whose views are different from our own?

To gain a comprehensive picture of the many views on death and the afterlife, I have benefited from the insights of Diana Eck, a professor of comparative religion and the head of the Pluralism Project at Harvard University. In her work, she recognizes that the diversity of cultures is a demographic fact. Pluralism, on the other hand, allows us to celebrate the differences in an effort to find our own deeper truths and paths of meaning. Rather than asserting the ultimate reality of one system over another, pluralism encourages us to view our beliefs as interpretative frameworks that can honor other frameworks. In our interconnected world, Eck compellingly argues, we need new forms of thought and noncontentious discourse that can help us to find the passageways between diverse worldviews and faith perspectives. As she said in her Gifford Lecture at the University of Edinburgh in 2009, such moments will occur through our points of connection "not in our libraries and studies, but in relationships with people of other faiths where we learn to speak and listen anew within earshot of one another."[1]

As stated in the worldview transformation model, engaging with different worldviews can lead to fundamental shifts in understanding our own experiences and insights. It can reduce dogmatic assertions and invite greater humility and compassion as we move to hold our understandings—and those of others—with a newfound curiosity and openness. Learning about other people's truth systems offers what Eck refers to as the "seedbed of a cross-ontological truth." The germination of new perspectives in a common ground that includes many expressions of lived experience allows us to deepen and expand our own individual worldviews and transformative practices. It leads us toward an ecology of perceptions that creates a fertile context for growth and discovery. Bringing greater awareness to our own mortality and our beliefs about an afterlife can impact our psyches by inviting new ways of being together in the world. It can also suggest new ways of understanding what may come after we die and how we can find peace in this understanding. We may harness the full intelligence of our minds and the wisdom of our souls to gain a universal sense of our oneness and our shared human experiences. This process may reveal more aspects or dimensions of life than we had previously observed.

## GLEANINGS

In this chapter, we have encountered a cross section of different worldviews on death and the afterlife. We have heard the voices of people who bring together both old and new perspectives, helping shape an emerging discourse that honors pluralism. Each voice that we heard speaks to the ways in which we can appreciate the perspectives of diverse traditions, while staying true to our own worldviews as we grow and transform. We heard from many people who do not fear death. Rather, their death awareness informs how they live with purpose and meaning.

The world's traditions offer many ways to understand and engage with death. As we examine the diversity of perspectives, we are invited to reflect on our own worldview about what it means to be fully human, now and beyond this lifetime. By examining our own beliefs and assumptions, we may begin to bring the fear of death out from the shadows of denial and into our lived discourse. In later chapters, we will explore in more depth some of the transformative practices that are used to overcome the fear of death and transform grief into life-affirming actions.

Bringing death awareness to the foreground in the context of pluralism invites us to broaden our in-group to include a more expansive sense of the "we," as predicted in the worldview transformation model. This expansion may offer new possibilities for living fully in our complex and multicultural world. The great work of our time includes fostering the emergence of a new worldview that is large enough to encompass the various possibilities and ways of engaging truth. Informing this new inclusivity is the truth system of science. In the next chapter, we will examine the evidence for consciousness beyond the brain and body. Exploring the scientific evidence for life after death brings us into the meeting place of noetic ways of knowing and the objective pursuit of reality.

‹ PRACTICE ›
## Shifting from the "Me" to "We"

Find a comfortable place to relax. Reflect on your own curiosity about the world. Consider the way in which you question worldviews that are different from your own.

As you do this, think about a time when you had a positive experience with someone whose worldview was fundamentally different from your own. Maybe it happened when you were traveling and were in another culture. Perhaps it was closer to home, when you met someone from a different race or ethnicity. Allow the image of this encounter to become a mirror to see who you are.

In what ways was this person like you? How were they different? Can you find areas of common purpose that give you new insights into your own worldview? Did you learn anything about the person or yourself that surprised you? What images and thoughts come up for you? What specific memories, sensations, and experiences arise in the moment? Journal on these, or any insights you may have had from this chapter, for at least ten minutes.

Consider sharing this practice with a friend or family member and discuss what you have each observed. Allow time for deep listening to each other before engaging in discussion about what you seen as common to your experiences.

# Chapter 5

# SCIENCE OF THE AFTERLIFE

With regards to end-of-life experiences,
it became perfectly clear to me, when we
studied cardiac arrests and near-death
experiences, that if people were right, then
this would be a very good model for the
death process itself.

PETER FENWICK, MD

The mainstream scientific worldview is grounded in materialism. The idea that our consciousness may extend beyond the body is considered by many scientists to be heresy. Yet a small cadre of trans-disciplinary scientists are exploring the thorny existential question that has filled humanity's imagination throughout time: What happens when we die? Such post-materialist scientists are approaching the topic of consciousness and the survival of bodily death from many perspectives that move beyond a strictly materialist science.

Putting together the scientific case for postmortem survival is a bit like assembling a jigsaw puzzle. Piecing together the evidence involves the kind of naturalist science conducted by Charles Darwin as he sought to document evolution through the use of diverse data sets. Studies of death, near-death and out-of-body experiences, reincarnation, and nonlocal consciousness all have a part in the emerging picture that lies before us. Informing this conversation are the exciting developments in science that bring new tools to ancient questions—and new questions to unchallenged assumptions.

## MAPPING DEATH EXPERIENCES
## AND BEYOND

During his academic career, Peter Fenwick was a neuropsychiatrist at the Maudsley Hospital in London, where he ran the neuropsychiatry-epilepsy unit for years. Fenwick has published myriad mainstream science papers on brain function. Less mainstream is his passion for the study of death, near-death, and end-of-life experiences. He explained:

> I was led to this work really by chance. I believed
> that there was nothing worth studying in near-death
> experiences. They were experiences that happened in
> California. They never crossed the sea to England. I
> thought they probably were more imagination. Then a
> case came into my consulting room. The man had had
> a cardiac arrest. He'd had an astounding near-death
> experience, and I knew that they were real.

Fenwick's transformation came as he sought to build linkages between materialist and post-materialist science. It has been the data, rather than his own noetic experiences, that led him to reformulate his worldview. Many would consider him a maverick for exploring these hidden realms of mind. Following William James as an inspiration, Fenwick adopted the attitude of radical empiricism. In the late nineteenth century, working against the tide of growing behaviorism in psychology, James had argued that any domain of human experience is a valid topic for scientific investigation. Perhaps it involves the experience of a ghostly presence or an apparition. Maybe it's an out-of-body experience where people report that they are actually watching their physical being from a different point of view. Whether it's the sensory kinds of experiences that many people describe, including music and harmony, or mystical states, anything can become a data point that we can examine across different explanatory systems—both noetic and rational ways of knowing.

Like James a century earlier, Fenwick sought out people who had experienced special states of consciousness. He listened to many descriptions of transcendent and ecstatic states, and he has looked for the patterns that connect diverse reports. His early career interests led him to read widely in the Eastern mystical literature, as well as in the hard sciences. He explained to me that "our day-to-day conscious experience is only a fraction of what is possible. As yet, we have no clear understanding either of the nature of consciousness or of its range."

The unconventional neuropsychiatrist has documented people's experiences at the moment of death. Using his scientific discipline, he has collected reports in which dying people reported insights into life after death. His goal has been to chart what happens to people when death comes and consciousness disintegrates. Many of Fenwick's patients report having otherworldly guides—such as departed relatives or friends—who come to help usher them into the spirit realm. Fenwick has also reported on those who seem to witness apparitions. This involves people "not only feeling like they are communicating with somebody, but actually having an experience of some kind of energy."

Such experiences, according to Fenwick, suggest that people may have access to alternate realms of reality. In 1987, he and his colleagues created a documentary film, *Glimpses of Death,* about near-death experiences. The researchers had received over two thousand letters from people about their encounters with other realms. They selected five hundred of these people and administered a questionnaire that gave the researchers insights on which to grow future work in hospices and nursing homes throughout Europe.

Fenwick is convinced, after his lifetime of delving into these extended realms of consciousness, that science can be useful in understanding these experiences of an afterlife. In near-death experiences, for example,

> we know perfectly well what happens because we have
> very good accounts from those who have them. We are
> also able to tell what state the brain is in, particularly
> if you use cardiac arrests, because there is no brain
> function. So science can understand that.

Then you come to the edge of science. You have something else which science, so far, has not examined properly—and in some cases, not examined at all. How is it that these people can have experiences when their brain isn't functioning? We need much more clarity on that. If it's true that you really do have experiences when the brain is not working, then it means that consciousness, or mind, is, in fact, not the same as brain. It's a fundamental step.

There are close similarities between people who have near-death experiences and those who are dying, according to Fenwick. From his case studies, he reports that some people who are dying or who have been near death say that dead relatives come to meet them and help them through the dying process. During this transitional phase, some people seem to transit in and out of another reality, in ways that share the same characteristics as those seen in the near-death experience. Fenwick continued:

Around the time of death, shapes are seen leaving the body, and the room is sometimes illuminated by a spiritual light, which is not dissimilar from that seen in the [near-death experience]. One of the most interesting aspects of the approaching-death experiences is the deathbed coincidences which are reported. These show a linking between people who are emotionally connected even though they may be continents apart, suggesting that mind has a nonlocal quality.

How does Fenwick, a hard-core neuroscientist, explain the phenomena he is reporting?

"It's going to be a difficult one to fit into current science," he readily admitted. "It looks as if, when brain function is down, there is a set of experiences in which you leave your body and sometimes even watch the resuscitation process. Now, what happens during this out-of-body process? No brain function. You've got a cardiac arrest. You're not breathing. All the brain stem reflexes have gone. So it is, in fact, a very good model for death itself."

From here, Fenwick considers the nature of the mind. He raises the question of whether the mind can function if the brain does not. Perhaps the mind works differently than the brain, he speculated.

> For example, it seems to be able to do things that you and I can't do in our normal minds. If you are up on the ceiling, which many are after a cardiac arrest (at least consciousness is), then when they look down, they can see the top of an object. We all could if we were up there, but what we could never see is the bottom of the object as well. They can see all sides of it.
>
> It's as if they have a multidimensional view. Some of them say they can see 360 degrees—again, a multidimensional view. So it looks as if you have to argue that in that state, we are multidimensional beings in a way that we're not when we are in the normal state of consciousness. It's as if a veil of some sort is removed. Then, of course, you go with the rest of the experience. The model for that would be that, somehow or other, there is in fact a transitory period into another dimension before the death process is complete.

As a physician and scientist, Fenwick has been part of many people's transition into death. Exploring these end-of-life experiences, he reports that people seem to move in and out of nonordinary realities.

> That links up very nicely with the near-death experience. And then at the time of death itself, you get a most amazing number of phenomena which I link to the loosening of consciousness in some way. Light can surround the body. The person who's dying can go and visit somebody. And then you get things happening in the room. Television switches off, the alarms go on. All that sort of thing occurs sometimes

when people die. So it's a very complex thing. It's not just a switching off. It can be. But on the whole, there are circumstances where you get these large number of phenomena which do occur.

Fenwick illustrates his points by citing the case of a person who "has gone the furthest." The individual was an air traffic controller who had a stable life. Then he had a cardiac arrest. His experience involved classic dimensions of a near-death experience, including a tunnel of light, encounters with a light being, and the entry into a room where he acquired information. Fenwick explained:

> He got to a point where he was being asked questions, and he knew the answers to all the questions. The knowledge he gained was about the structure of the universe and so on. He then left the room and started to transform into energy. He describes himself as floating, as pure energy. Then he was moving towards a cosmic source of energy with which he was going to fuse. And remember, when we say "energy," we're talking about love, light, and compassion. That's what it's composed of. As he was coming up to fusion with this, he had realized by this time that we never die. In fact, death is simply a transition process from our life here to experience after death.
>
> He had to go back and tell his wife. And this act on his part took him right back into his body. He's very clear, he says, that what happens after death is very moral. You have to face what you've done. Nobody churches you, but you church yourself. He's also very clear that the whole of the universe is just one thing. So the model then is that there are filters. And here, I'm with William James on this, that the filters seem to stop us seeing much of what is beyond. But in certain circumstances, the filters break down, and then you get a glimpse of what actually is. And that is what he seems to have had.

While not every case is the same, the air traffic controller had "death-bed visitors." Such visitors or spirit beings occur in 25 percent of the patients studied by Fenwick.

> Then at the bottom end, you get [visits from] siblings
> and wives and husbands. And in our culture, in our
> series, only three percent [were visits from] angels. So
> those are the people who come. Many times the person
> is so close to dying that they [the visitors] can greet
> them, smile at them, welcome them. They [the dying
> people] can't say anything because they're too weak.

These cases of deathbed and near-death experiences suggest to Fenwick that consciousness extends beyond the body. "The best evidence is the fact that consciousness in these situations appears not to terminate. It appears to go on," he said. "But then you have to ask the question, if it goes on, what is its form?"

Fenwick shared with me his own personal worldview about the survival of consciousness and the inevitable loss of his own physical body—a worldview based on his many years of experience. His research has offered him a sense of confidence that is free of fear.

> I'm absolutely certain that there's no termination
> of consciousness. It is not a snuffing out. It is a
> progression into another form of reality. The accounts
> we've had of this reality mean it's rather like this one
> now. But what people can do, and how they move,
> and so on, is different. But it looks as if there's a
> continuation of personal consciousness in some form.
>
> Do I fear death? Not at all. I've seen far too many
> near-death experiences to fear death. I've also seen far
> too many end-of-life experiences to fear death.
>
> Now, do people fear death? Yes. Why? Because
> they haven't done the process of dying. Most people
> are afraid of the process of dying, rather than the
> switching off. But if they looked at the data, their

fear of death would go. If you accept that the near-death experience, particularly in cardiac arrest, is a good model for dying, you have nothing to fear at all. You're looked after the whole way. So when your time comes, enjoy.

## ADVANCES IN THE NEUROSCIENCES

Key to the science of consciousness in modern science is the functioning of the brain. On one hand, the physicalist worldview reduces all experience to neural firings. Fenwick, on the other hand, believes consciousness may extend beyond the brain.

> It's not really the brain itself that's exciting, although of course the wonderful complexity of the brain's mechanism and architecture are totally fascinating. But for me, the really exciting fact is that it appears to be the gateway to consciousness.
>
> Current neuroimaging, both by the examination of blood flow, as in fMRI scanning, and magnetically, as in magnetoencephalography, have shown us the physical structures which underpin mind and their relationship to function. The latest 9 Tesla fMRI magnets which are now coming on-stream will give us a resolution of brain function right down to a fraction of a millimeter, while magnetoencephalography will give us a time resolution right down to a fraction of a millisecond. With these enhanced resolutions, I suspect we shall find even more details of the mechanisms which manifest consciousness. Still, consciousness itself will be elusive.

As Fenwick suggests, contemporary scientists, steeped in new tools and technology, are equipped with new explanatory models that divide the lines between materialist science and myriad religious beliefs and practices. Positron emission tomography (PET) scans, functional MRI, and advanced multichannel EEG each offer a glimpse into the complex neural firings in our brains. With increasing precision, scientists in the twenty-first century are mapping emotions and attributions of intention, and even how they believe mystical encounters with other worlds, near-death experiences, and beliefs in reincarnation can be explained by physiological circuitry reduced to a place inside our heads.

Most scientists do not believe that near-death experiences are evidence of an afterlife, as psychologists Dean Mobbs and Caroline Watt reported in an article in the scientific journal *Trends in Cognitive Science*.

> There is nothing paranormal about near-death experiences: how neuroscience can explain seeing bright lights, meeting the dead, or being convinced you are one of them. . . .
>
> Taken together, the scientific evidence suggests that all aspects of the near-death experience have a neurophysiological or psychological basis: the vivid pleasure frequently experienced in near-death experiences may be the result of fear-elicited opioid release, while the life review and REM components of the near-death experience could be attributed to the action of the locus coeruleus–noradrenaline system. Out-of-body experiences and feelings of disconnection with the physical body could arise because of a breakdown in multisensory processes, and the bright lights and tunneling could be the result of a peripheral to fovea breakdown of the visual system through oxygen deprivation. A priori expectations, where the individual makes sense of the situation by believing they will experience the archetypal near-death experience package, may also play a crucial role.[1]
>
> Several neuroscientists today are blazing new trails in understanding the neurological processes underlying a near-death or out-of-body experience. Cutting-edge

research is now underway in Switzerland, for example, to build a materialist argument for these mystical states of consciousness. Here, work is being done to induce these out-of-body-type experiences through an electric shock in a portion of the brain called the angular gyrus. One woman, for example, reported having the sensation of hanging from the ceiling, looking down, and seeing her body, in a way very much like other people have described the sense of their experience beyond the body. Studies of the brain suggest to these Swiss-based scientists that "activation of these regions is the neural correlate of the disembodiment that is part of the out-of-body experience."[2]

But do these studies provide evidence of consciousness beyond the body? While materialist explanations that reduce mysticism exclusively to the brain are compelling, many post-materialist scientists argue that they are simply incomplete. As Fenwick and others will quickly point out, this worldview does not actually explain the experiences that people have reported. Correlates of brain activity are not the same as causation. Electrical activities in the brain are not the experiences themselves.

Still, studying the brain offers many insights into consciousness beyond the brain. The neurosciences have developed vast new technologies and approaches for understanding consciousness. Expanded research programs include the possibility of expanded reaches of consciousness beyond the body. These studies have important implications for the possible survival of consciousness after bodily death.

## MEDIUMSHIP AND COMMUNICATION WITH THE DEPARTED

To address this matter of consciousness head-on, several scientists interested in the afterlife hypothesis are studying mediumship. They are focusing on people who report direct communication with the deceased and evaluating the accuracy of these people's communications.

They are also exploring what is going on in these people's brains and bodies when they claim to be speaking with the deceased.

Julie Beischel received her doctorate in pharmacology and toxicology, with a minor in microbiology and immunology, from the University of Arizona. She is, by all academic standards, a mainstream scientist. Today she is conducting research with people who purportedly talk with the dead. Working with her husband, Beischel runs the Windbridge Institute for Applied Research in Human Potential in Tucson, Arizona. They focus on the experiences of and information reported by a team of mediums. Beischel shared with me her own worldview:

> When I think about my death, I agree with what
> Woody Allen said, "I'm not scared of death. I just
> don't want to be there when it happens." I have
> fear about dying, but I don't have fear about being
> dead. It's my understanding that survival is just a
> characteristic of consciousness. It's just what happens.
> When you die, it's a different sort of existence, and
> just like when you came into this body, you had to
> learn how to use this body and move its parts. You
> have to learn how to not have a body anymore and
> how to exist that way. So as a scientist, I sort of look
> forward to that experience of what it feels like to be a
> mind without a body.

Most strict materialist scientists dismiss the very idea that people can conduct serious research on mediumship and spirit communication. To address the methodological challenges, post-materialist scientists make use of elaborate procedures and rigorous scientific controls. Their goal is to bring an objective approach to the study of subjective phenomena that lie far beyond mainstream materialism.

Two innovative studies with mediums demonstrate this objective approach. Arnaud Delorme is an experienced psychophysiologist who created EEGLAB, an open-source toolbox for analyzing brain-wave activity. Having established a sturdy foundation for his professional credibility and expertise, he decided to follow up on a personal encounter he once had with a medium. A direct noetic experience catalyzed his own transformative process, as predicted by the world-view transformation model, and led him to shift his career interest.

Working with Beischel and others, Delorme conducted his two studies at the Institute of Noetic Sciences. The researchers recruited four mediums who had previously worked with Beischel at her center in Arizona. The mediums were chosen because they had already shown encouraging results in possible spirit communications under various scientific conditions, according to Beischel. As Delorme explained, the mediums were a diverse group.

> They were really very interesting people. . . . They were all different types. The only man was a business medium who was managing other mediums. We had a super-cautious medium who was like, "Okay, I have a reading tomorrow, and people are paying a lot of money. I'm going to do my two-hour meditation now. . . . I have to isolate myself to be in the best possible state." We had a medium who was excited about the research and absolutely wanted to convince us that it was worthwhile. She was doing readings of everybody in the room as we were having lunch. . . . All of them had experiences in their childhood or some type of history [of] their ancestors doing similar things.

The scientists designed the first study to see if the mediums' brains functioned in unique ways when they felt they were in touch with a departed spirit. The researchers compared sessions involving ostensible spirit communication with times when the mediums were instructed to use their imagination—to make up a story in their head or to think about a relative. Their brain activity and frequencies were then measured to see if there were objective changes that correlated with the mediums' subjective experiences. Electrodes were placed on the scalps of the mediums to record electro-cortical activity. Investigators then measured the mediums' physiology during four distinct mental states: recollection, perception, fabrication, and communication with the deceased.

According to Delorme, the scientists were able to show statistically significant accuracy measures for the mediumship readings of three out of four mediums. The electro-cortical activity results showed significant differences among all four of the induced mental states. These differences suggest to the scientists that mediumship may involve a unique subjective experience that is different from the mental activity associated with imagination or memory recall. In particular, when the mediums were cavorting with spirits, their state of mind involved a slowing of theta brain-wave activity, which in turn correlated with higher accuracy in their spirit readings. The scientists' interpretation is that the emptier the brain is in terms of working memory, the easier it is for the mediums to gain reliable information about the deceased person.

In a second study, the team set out to measure the brains of the mediums when what they said about spirits was judged to be accurate. They had each medium do a reading of a departed person, who was identified as the "target dead person" in the study. It was the job of the research team to determine how closely the reading matched what the researchers knew about the target dead person. To avoid bias or wishful thinking, the researchers asked the "sitter" (the person for whom the reading was being done) to evaluate the reading done for the deceased person they knew (the target dead person), as well as a reading from another deceased person whom they did not know (the control dead person), without knowing which readings were which. The sitter was instructed to evaluate each statement made about the two dead people, line by line. Making use of a complex statistical process, the researchers were able to evaluate the overall accuracy of the readings against one another. They found statistically significant evidence that the readings were more accurate for the target dead person than for the control dead person.

Do these two studies prove the existence of life after death? Delorme hedges his bets; he notes that these are preliminary studies. There may be some unseen flaw, challenges of replication, or simply not enough data. It is also possible that the results may be attributed to a phenomenon known as "super psi," which means that the study results, while valid, may reflect telepathy among the living as compared to communication with the deceased.

Dean Radin, whom we met in chapter 3, was a member of the team who worked with mediums at the Institute of Noetic Sciences. He described how difficult it is to reconcile experiences that people report in everyday life with the kind of controlled, laboratory research that scientists like to do.

> In order to answer the question of whether consciousness transcends the body, we need to know what consciousness is. And we don't know what consciousness is. What I believe, though, based on the scientific studies, is that there's some aspect of awareness that extends beyond the body.
>
> Now, that doesn't necessarily mean that consciousness can become detached. It's like something that is separate from you somehow, or you are it, and you're inside it—I don't think we actually know enough yet. And not only that, but the evidence from parapsychological studies doesn't actually suggest that consciousness is separate from the body. When we talk about a mind-body connection, it may be two sides of the same coin, in which case you need a body in order to sustain consciousness, and maybe vice versa.

The super-psi hypothesis is one that is championed by Daryl J. Bem, whom we met earlier. He has spent decades studying psi (or psychic) phenomena and providing compelling evidence for experiences such as precognition—knowledge of the future.[3] Bem acknowledges his own worldview to be complex and dynamic.

> I have a sort of split brain on where I am in this whole issue of materialism versus not. I've read the research on the afterlife. I am still not persuaded by that because I'm open to the possibility that we're dealing

with psi, clairvoyance, telepathy, among the living.
. . . I resist the notion that we're seeing the afterlife
or essentially communicating with people who have
departed. But I'm certainly open to the possibility that,
as a source of energy, it [consciousness] doesn't just
vanish with our bodies.

This kind of research makes me believe that
our consciousness extends well beyond our own
bodies and our own brains. So having an extended
consciousness of some sort accords quite well with
what modern physics is trying to deal with. The
nature of time is still an open question in physics. . . .
Most physicists agree that the actual laws of physics
are time-symmetric: they do not distinguish between
time flowing forwards and time flowing backwards.

Not all scientists accept these results from the psi research or medi-umship. In fact, many are highly skeptical of both psi data and studies of consciousness beyond death. Bem, a social psychologist, speculates about why this may be so.

Many scientists are opposed to this because they
are afraid of looking foolish, and they don't want
to make a mistake. And what they overlook, which
they should know from their statistics course, is you
can't avoid making a mistake. You have a choice
of what mistake to make. A type-one error means
you draw a conclusion that there's something there
when there isn't. [In this field of research] that's what
psychologists are really afraid of because that will
make them look foolish. Adopting very strict criteria
before you're willing to say that something is there is
what's called the type-two error. The type-two error
is that you overlook something that is there because
you are so skeptical and demand such a high level of
confidence before you're willing to commit yourself.

At the end of the day, objectivity may be hard to achieve. Whatever the mechanism by which the mediums obtained information about the deceased people, Delorme has approached these topics with an interest in his own consciousness transformation. Learning new things through the lens of science may help broaden his understanding of the world in which we live. He told me:

> The science is very cold. Data. Statistics. It's always good when we have these mediums to discuss with them, get more live interactions, do readings, tell stories. . . . Even if mediumship is not real, it has been shown to be beneficial. It helps individuals to move forward.
>
> I think the data from these studies of mediums tells us something about consciousness that we don't accept right now in the world. Gaining information that is beyond sensory perception tells us about the nonlocality of consciousness and extended properties of consciousness that we think are not there. . . . I'm strongly anchored in the Western scientific model, where science holds the truth. If I figure out that I have proof, at least beyond any doubt, "Okay, there is a nonlocal aspect to consciousness, or survival of consciousness," there might be a click in my mind that allows me to go to the next stage. Just because my mind is so anchored in science, maybe my whole model will change—evolve. It hasn't happened yet. We'll see.

## DOCUMENTING EVIDENCE FOR REINCARNATION

A central question related to life after death is fundamental: What, if anything, survives? Scientists have aimed to answer to this question by studying past lives.

Ian Stevenson spent decades collecting case studies that are suggestive of reincarnation. Prior to his death in 2007, Stevenson was a senior psychiatrist and chairman of the Department of Psychiatry at the University of Virginia. When he heard reports of young children from various parts of the world who were saying they had memories of a previous life, he decided to investigate. Never taking the cases at face value, Stevenson and his team set out to document apparent cases of reincarnation. His method was to examine what the child said and compare it to the biographical profile of a person who actually lived and died before the child was born. While the methods were not well-controlled laboratory studies, Stevenson and his team approached their studies with rigor and caution.

The methods that the researchers used varied over time and from case to case. Generally, Stevenson would hear about a case of a child talking about a previous life. He would then go and talk with the family, documenting as carefully and exactly as possible what the child said. If the person from the previous life had already been identified, then the researchers focused on what the child had said before that identification had been made. In other words, they focused on what the child had said before the child's family knew anything about the person that the child was talking about. Stevenson then went to the other side of the case and talked to the previous person's family. With painstaking care, the psychiatrist went through every statement that the child made to evaluate just how accurate it was.

Some of the most provocative cases that support the reincarnation hypothesis involve birthmarks and birth defects. In such cases, a child was born with body signs that matched the wounds, usually fatal wounds, on the body of the previous person. The scientists sought out autopsy reports and interviewed witnesses who had seen the body. Eventually Stevenson published a book, *Reincarnation and Biology: A Contribution to the Etiology of Birthmarks and Birth Defects*, describing over two hundred of the best cases that he felt were suggestive of reincarnation.[4]

Jim Tucker, MD, is an associate professor of psychiatry and neurobehavioral sciences at the University of Virginia. He and other researchers are continuing the work of Ian Stevenson at the UVA School of Medicine's Division of Perceptual Studies. Tucker was in private practice as a child psychiatrist for nine years before becoming intrigued by the research findings on the possible survival of consciousness. He gave up his practice to focus on this topic. Remarriage and a life transition opened Tucker up to spiritual and psychic phenomena, catalyzing a worldview transformation, yet he maintained a critical mind. Understanding reincarnation was not a matter of faith for him.

During an interview, Tucker acknowledged that mainstream medicine focuses on extending and improving life, but members of the health profession are largely reluctant to explore the question of life after death. At UVA, he feels supported by a team of researchers who are doing "very careful, methodical, and thoughtful work on this big question that all of us wonder about."

Based on this research, Tucker is convinced that there is more than just the physical universe. The study of reincarnation may help to expand our understanding of reality. Like other post-materialist scientists, Tucker is especially intrigued by the idea that consciousness may be what survives death. In the case of reincarnation, the evidence from case studies also implies that identity can survive the death of the body. Tucker explained:

> There is this consciousness piece which I think is a
> separate entity. It may well be the primary entity that
> the physical world grows out of. That raises all sorts
> of questions, including whether the consciousness in
> each of us is created by our brains, or if the brain is a
> conduit for it, and the consciousness can continue on
> after the brain and the body die.

Having evaluated the data supporting reincarnation, he noted that many of the twenty-five hundred cases collected by Stevenson and his team are evidential. The best of these cases have no easy materialist scientific explanations. Tucker described two case studies from his own data in America.

> One is a little boy named Sam Taylor who was born
> eighteen months after his father's father had died.

When he was about a year and a half old, his dad was changing his diaper one day, and Sam looked up at him and said, "When I was your age, I used to change your diapers." His parents were thrown by that. They certainly never considered reincarnation. In fact, his mom was the daughter of a Baptist minister. But he kept saying this stuff—kept saying, "I was Grandpa, and I used to be big."

His parents became intrigued by that. His mom would ask him questions, and he came up with some pretty interesting details. He talked about the grandfather's sister being murdered. In fact, she had been killed some sixty years before. His parents felt certain that he'd never heard about it. He also talked about his wife from the past life, who in this life was his grandmother . . . making milkshakes for him every day at the end of his life. And not just that she made milkshakes, but that she used a food processor to make them instead of the blender. He said some specific details.

She died when he was about four and a half. His dad went out and collected belongings and came back with some family photos. Sam's family had not had photos of his dad's family in the home. So one night, Sam's mom spread them out on the coffee table, just looking at the various pictures. Sam walked over and started pointing to pictures of his grandfather and saying, "That's me, that's me." There was also a picture of his grandfather's first car. There's no one in the car, just a picture of the car. He got very excited and said, "Hey, that's my car." To test him, his mom showed him a picture of a class photo from elementary school and said, "Okay, show me where you are in the picture." He ran his finger along the different faces and stopped on the one of his grandfather and said, "That's me."

That was one that adds what we call recognition, where he was able to pick out his grandfather in a group photo. It's also what we call a "same family case" where a child seems to remember the life of a family member. Those cases have the inherent weakness in that you wonder if they could have picked up information through normal means. But then, in those cases, children might come up with details the parents felt certain they did not overhear, and with the recognition, that makes it a stronger case.

The second case was one Tucker and Stevenson had investigated together. It offers an example of biological evidence supporting reincarnation. Such biological markers are perhaps the strongest data supporting the reincarnation hypothesis because they are physical in nature.

The child was born with three birthmarks that seemed to match lesions on his deceased half-brother. This little boy was named Patrick. His half-brother, Kevin, when he was about a year and a half old, started limping, and he eventually fell and had what is called a pathological fracture of his leg. He went in for medical work that included a biopsy of what turned out to be a tumor above his right ear. He was diagnosed with metastatic cancer. One of his eyes was bruised and bulging. For the treatment, they put a large IV into the right side of his neck and ran the chemo in through that. The side became inflamed at times, but basically he tolerated the treatment fairly well.

They went home, but then [the cancer] came back six months later. At that point, his mom reports that he was essentially blind in his left eye and was having bleeding because [the cancer] had infiltrated his bone marrow. Anyway, they did a day of radiation, but he went home and died a couple days later.

His mom was devastated by the loss. Eventually, she and Kevin's father split up. She remarried and had a girl, another son, and then twelve years after Kevin died, she had her third son, Patrick.

She soon noticed that he had things that matched what Kevin had had. His left eye was covered by an opacity, what was diagnosed as a corneal leukemia. That essentially made him blind in that eye the way that Kevin had been. He had a nodule over his right ear, which matched the biopsy in Kevin's case. He had a mark on his neck where the [IV] had been in Kevin. It basically looked like a small cut; it was this little slanting line. When he started to walk, he walked with a limp, favoring that leg just like Kevin had done even though there was no apparent medical reason for the limp. Once he got old enough to talk, he talked about several things from Kevin's life, including accurate descriptions of the home where Kevin and his mom had lived at the time and various incidents that took place.

We were able to get the medical records and document the various things that his mom had told us about Kevin, including one handwritten note that showed the mark on his neck was on the same side where Patrick's birthmark was. . . . It was quite comforting for his mother to think that she had, in some ways, had Kevin back again. But that doesn't mean that you dismiss the case. Sure, she was happy with that, but there were also these other features that warrant attention.

There are many materialist scientists who maintain deep skepticism about these data because they don't fit within the materialist model. Those who are critical of the findings for an afterlife use words like *anecdote, coincidence,* or even *delusion.* As Tucker noted, alternative explanations have been offered for these case collections. First, it has been asserted that they represent coincidences that have no scientific value. Many people have lived and died, so some details are going to correspond just on the basis of chance. Tucker explained:

> Some people criticize the work simply by dismissing it out of hand, saying that reincarnation can't happen, so obviously there's no value to this work. But that's really not a very scientific—sort of scientism perhaps, but it's not a scientific view, if you let your beliefs prejudice your view of the evidence when it should, of course, be the other way around.

Summarizing the nature of the evidence, Tucker reported:

> After fifty years of research, now we've got twenty-five hundred cases in our files of young children who reported memories of previous lives. Some of them occur in places with a belief in reincarnation, but others in places like the US without a general belief in reincarnation and without a family belief in reincarnation.
>
> Several hundred of the cases have involved birthmarks or birth defects that match wounds on the body of the previous person. And some of them have also included written records where we know precisely what the child said about the previous life. We can match that with the previous person who has been identified. And if you take all these cases as a group, what you're left with is significant evidence that there's not a normal explanation that easily can discount these cases.

How do we know what we know? What amount of data can convince us to shift our worldview? For clinical psychologist Rick Hanson, like Tucker, there are enough examples of reincarnation to make him feel confident:

Even if we eliminate 99 percent of the anecdotal cases in which people report reincarnation—or people who report near-death experiences when they come back, or they seem to have some kind of contact with essences, or influences, or factors, or even personalities that have gone beyond—even if we eliminate 99 percent of that, from a mathematic standpoint it's an existence proof: all you need is one. It's just one case. It's difficult to imagine that every single one of these accounts is dismissible. All that gives me some sense that there is some kind of process that occurs after the death of the physical body.

## GLEANINGS

Evaluating the evidence for an afterlife is both an objective and a subjective experience. In this chapter, we have explored how science is approaching these questions. Though the study of life after death is still in its most nascent phase, important patterns are emerging. But perhaps the dividing line between science and spiritual belief is not as clear as some would think. Rather than clear-cut boundaries between mind and matter, a new and more fluid pattern is emerging that blurs the line between physical and metaphysical assumptions, like children's feet blurring the chalk on the sidewalk during a busy day of hopscotch.

While the data from near-death experiences may or may not be evidence that there is a consciousness that survives physical death, these experiences can have life-changing consequences. They are catalysts for worldview transformations. As William James alluded to many years ago, these noetic experiences can carry with them a curious sense of authority about the nature of life and what comes after. For researchers Arnaud Delorme, Peter Fenwick, and Daryl J. Bem, science is a transformative practice that helps them to better understand what is true. However, as Bem explained, in the nature of science, there are no absolute answers to any truth claim. Much of what we may know is based on our assumptions and direct personal experiences. In the following chapter, we will further consider various death awareness practices to help us transform our relationship to death.

‹ PRACTICE ›
## Contrasting Subjective and Objective Ways of Knowing

Take a few moments to center yourself and relax. Consider your own worldview and beliefs about death and the afterlife. How do you know what you know? What shapes your own evaluation of the evidence for an afterlife? How do you form your own opinions and beliefs? To what extent are they based on your direct noetic experiences? What role does science and its focus on objective knowing play in shaping your personal worldview?

Then, in your journal, draw a circle that fills the page. Inside, create a pie chart that is divided to show your own evaluation of what informs your worldview and in what proportion. Drawing in pencil will allow you to shift the proportions if needed. When you are done, take time to write for ten minutes about what you learned from this experience—about what shapes your worldview, including both subjective and objective ways of knowing.

# Chapter 6

# THE PRACTICE OF DYING

Being in touch with awareness, as opposed
to being in touch with the contents, even if
the awareness isn't something that survives
in its present form . . . [is a] stance [that]
makes you better able to deal with what
might be coming in those moments of death.

CASSANDRA VIETEN

There are many stories of people who have awareness and volition in their dying process. For Satish Kumar, being in the presence of his mother's intentional passing has helped him understand death. When his mother, who lived in India, was eighty, she decided it was time to make her transition. She told her family, "Now this body's too frail. I can't see properly, I can't cook properly, I can't garden properly, I can't walk properly. I want new life."

The family's belief in reincarnation empowered the elderly lady. She went to her children and said good-bye to everybody. When they asked her where she was going, she told them that it was time for her transformation. She decided to fast until death, beginning the next day. She was ready for a new body and wanted to begin the process. As the news spread, people came to visit with her, to sing songs and chant hymns and mantras.

"It was an atmosphere of celebration of her life," Kumar told me. "And for thirty-five days she fasted. Can you imagine? The celebration went on for thirty-five days! She was not afraid because she was aware that only the body was changing form. Life force is continuous, unfolding, evolving, but not dying."

The experience affirmed for Kumar that death should be at home, when possible, and in the company of loved ones. "The state, the community, the family, they should all support death in the arms of the family members," he professed. He then described another family member, his mother-in-law, who died at the age of ninety-four. Her death was also a powerful and transformative experience—and a gift to his entire family.

> When she was ill, my wife, my mother-in-law's daughter, and her grandchildren, my children, and a great-grandchild, they were all with her at home. And one evening, she said, "I feel a bit thirsty." So my daughter brought a glass of water. However, [my mother-in-law] could not sit up properly, so my son held her in his arms. And as she was drinking water, she could not drink, and she died in the arms of her grandchildren feeding her water. That is the best way to die.
>
> A new awareness is needed in the US: that one should give old people a death of dignity, a death of love. People should die in the arms of their children, grandchildren, great-grandchildren, or friends. But a lonely hospital death, that is not right. And America, such a wealthy country, such an advanced country, if in America one cannot give a dying person a death of dignity, which country can do it? A new movement of awareness, of death with dignity, is needed.

## THE IDEAL OF A GOOD DEATH

Western society has worked to define the pathways that lead to death. We know that the heart fails and the kidneys shut down. But for the most part, there has been little attention given to the detailed changes in the mental state of the dying or the ways in which we shift our consciousness or awareness.

As Janet Quinn, a nurse, educator, and teacher, noted, "With seven billion people on the planet, there are seven billion ways to die." However, there are forms and patterns that inform the practice of dying. The notion of a good death has its roots in various spiritual traditions. It becomes an obstacle that leads people to feel like they're not dying in the right way. Dying a good death means many different things to people. Being able to die where and how we want is important for many of us. Likewise, having adequate pain control and physical care is also something people desire. In addition, social and spiritual care are also parts of the dying process.

Many traditions tell us that our intentions regarding our death and the attention we give to dying can help inform the passage from life to death. By establishing how we intend our death to be and drawing our attention to death's transformative power, we can regain our own personal power in a society that has largely lost the practice of dying.

The great death transformation can be a period of reconciliation if it happens over time and not suddenly. The person facing their death can do so without guilt or regret, and family discord can be healed. Bringing awareness to end-of-life experiences can be comforting to the bereaved family members, as well as for the dying person themselves. Understanding death as a natural part of life removes the stigma and sense of failure and helps ameliorate fear and suffering.

As a member of the clergy, Lauren Artress has been part of a support team during the final hours in many people's lives. The idea of a good death is often not part of the conversation that clergy have with a dying person, even though it is part of the Christian tradition. Artress described her experience as canon pastor of Grace Cathedral during the height of the AIDS epidemic in the late 1980s, when she considered what it means to die a good death:

> We had ninety-two people die one year in the congregation. Living a good death or dying a good death (you could think of it either way) really means that the person dying comes face-to-face with that transition. And that they enter it as fully and as consciously as they can.

The quality of dying a good death is that the people who move through that transition with the person dying feel like they're given a gift. They may become more awakened, more enlightened; they may feel like they've had a once-in-a-lifetime spiritual experience. A gift is given upon departure.

Based in his own experience at the end of life, physician and cancer patient Lee Lipsenthal explained his own feelings as his final days were upon him.

For me personally, I want my family to see that I've died well. And I don't mean that as a mission. I mean that as I'm being me as I die; I'm not losing the core of me even though my body is changing and may be withering.

In the end, he achieved this goal. He was in the presence of loved ones as he passed gracefully.

## THE TIMING OF DEATH

Various sociological studies have shown that not all days are equal in terms of when we die. For example, it appears that adults who die have a greater chance of dying on Christmas Day, the day after Christmas, or New Year's Day than any other single day of the year. According to the Centers for Disease Control, this is true of people dying from the most common diseases: circulatory problems, respiratory diseases, endocrine/nutritional/metabolic problems, digestive diseases, and cancer. It is equally true if people die from natural causes. Interestingly, it does not apply to those who take their own lives.[1]

Mondays are another time that is correlated with death. A team of Scottish researchers reported that more people die from heart disease on the first day of the week than any other day. To date, there is no clear explanation for this phenomenon. Clearly, lifestyle and cultural factors are implicated; too much to drink, dread of work on Monday mornings, and stress may all take their toll on our physical wellbeing.[2] These death trends support the links between our bodies, minds, and social patterns.

And yet there are other factors that influence the timing of death. As was the case for Kumar's mother, death was a matter of intention. From his studies of at-death experiences, Fenwick reports:

> Some of the people who have described the appearance of deathbed visitors say they have come to take them on a journey. They say they've been able to negotiate a short postponement, perhaps because someone they want to say good-bye to is on their way to see them. It seems you may be able to manage a few days' extra time for some good reason.

## PREPARING FOR DEATH

Many religious and spiritual traditions have developed practices to help people prepare themselves for their earthly transition and, in doing so, transform their fear of death. The following are examples of three such transformative practices.

### Visualizing Death

Within Sufism, there is a tradition called *Melami*. This tradition of visualizing our own death focuses directly on the reality of death, yet also shows death to be a transition during which we are given the opportunity to awaken. I learned about this practice from Metin Bobaroğlu, an imam who lives in Turkey. Through an interpreter, the elderly teacher explained:

> We think that we came to this world for an experiencing of death itself, for preparing ourselves to die. . . . I want to explain about the practices of ancient Egypt, which we also follow. These are very complicated practices. But I will keep it simple.
>
> There are four degrees to it. The followers should experience some experiences. After this, they should

lay down in a grave, in a tomb, to experience death itself. After this experience, they will be awakened from this state. This is a ritual. You can find this ritual in all kinds of Sufi traditions. If we don't use this practice, then we cannot call ourselves Sufi. A Sufi is a man who dies before dying from natural death . . .

Our masters prepare us for this state. After this experience, they awaken us, and we can be ready for service. In [this] tradition, the people who don't die before natural death cannot be of service to people. This is the necessity of being a master, to have this experience.

This Sufi death preparation is a transformative practice, one that embodies the five steps of the worldview transformation model (see chapter 1). People hold the intention to move beyond fear in the face of death. Feelings that are festering in the unconscious are made explicit, shifting attention to the awakening that comes at the time of death. They engage in repetitive actions that allow them to live into a new way of understanding death based on the wisdom and methods of their tradition. In this process of mastery, the dervishes, or practitioners, are called to be of service to the larger community in a spirit of love that characterizes the Sufi path. As terror management theory suggests, learning to transcend the fear of death within a supportive community of fellow practitioners that fosters positive self-worth and the call to service may help people heal what Ernest Becker defined as existential terror.

Bobaroğlu described his own experience around the anticipation of death. Drawing specifically on the Sufi tradition of *Naqshbandi*, the Sufi death-awareness practice he described involves specific steps. Central to the practice is the use of imagination to connect with death. The followers also engage in fasting and the repetition of divine names. After these actions, they put a veil, representing a curtain, on their heads.

Then they begin to imagine that they are dead, that they are buried in a grave, that their body is being washed before the funeral. They imagine that their funeral is completed. Their death is realized, and there is the realization of this by the other people. The dervish should practice this imagination of his own death.

After people die, they continue to exist in a form of light body. When we leave our bodies, we continue to live in a state of a dream. This is an intermediate state. It is called purgatory, or maybe, isthmus. Prophet Mohammed says, "For this sleep is an example for death. And the dream is an example for the life after death."

The critical point of our tradition is to experience love, the divine love. First, when we are born from our mother with our body, we have a basic anxiety in us. This basic anxiety is because of death. To be born is a big trauma for the spirit itself. . . . That's why, in our tradition, the body means prison. And we wait for the experience to be free of this prison. To die is an experience of rebirth. To be dead, to die, is freedom itself.

The experience of death is something very real and common in all traditions. All traditions and all religions are the same in essence. If we can understand the essence of them, then we can live together in peace.

## Mental Rehearsal

Native American practitioner Tony Redhouse has developed his own practice, combining traditional and noetic forms of guidance to prepare for death. He mentally rehearses dying, with the intention of transforming his fear of death. Through a repetitive act of visualization, he seeks to bring his attention to those aspects of life he had been ignoring.

What I'm learning is this: I'm putting myself into the hospice bed in my life now because I'm seeing certain dynamics. When you're lying in a hospice bed, you are going to look at your whole life, everything that you created, and whether you have lived true to yourself or whether you have lived your whole life for the expectations of everybody else.

You're going to be at that point, and you're going to look back, and there's going to be some regret. What I've learned in this is that, right now, if I place myself in a hospice bed and if I think about what it is right now that I really want to fulfill, what dream do I want to fulfill in my life right now, then I have the ability, the passion, the strength, to do these things. I have the energy to do them. I'm going to do them right now because I'm not going to wait until I'm in a hospice bed and say, "You know what? I never did what I wanted to do."

Employing intention, attention, repetition, guidance, and acceptance of his place in the cycle of life—five steps of the worldview transformation model (see chapter 1)—is what makes Redhouse's mental rehearsal a transformative practice.

## Becoming Aware of Thoughts and Emotions

Cassandra Vieten is a clinical psychologist and president and CEO of the Institute of Noetic Sciences. She explained to me that one way of moving beyond the fear of death and into possibility involves monitoring our thoughts and emotions.

There are thoughts that are always moving through our minds, almost as though they are train cars that are attached to each other. They just keep chugging through all the time. . . . Sometimes you have a pain, a tension, a pleasure, or an infinite number of body sensations that one can have, but they're always temporary, and they're ever flowing.

Then there is the realm of emotions . . . and feelings, where you might have anger, happiness, sadness, joy, and then mood tones, like pressure or anxiety. These are all almost like weather patterns in the sky. Sometimes there are hurricanes and tornadoes and big storms and lightning and thunder, and they're really intense. Sometimes the sky is relatively clear, and you've got very subtle emotions and thoughts and feelings. Sometimes it's just like clouds that hang over the sky for days, like a mood. But the point is that the sky is the awareness that holds all of these emotions. . . .

When I have experienced that sky mind, or the awareness that is aware of all of the contents that move through experience that are temporary and are changing, there's almost a timelessness. And it makes me wonder whether that awareness is perhaps something that survives.

The practice is being in touch with awareness, as opposed to being in touch with the contents. Even if the awareness isn't something that survives in its present form, I think that stance makes you better able to deal with what might be coming in those moments of death.

Different meditation and contemplative practices can help us develop the ability to observe our thoughts and emotions, as Vieten described. Such practices can help us deal with fear and other sensations and thoughts we may have about death. Through mindful repetition, new habits can be developed, based on rich histories within various contemplative traditions. In this way, mindfulness can become a transformative practice to cultivate positive experiences surrounding death awareness.

## GLEANINGS

In this chapter, we explored instances in which preparing for death can transform our fear of death. Such transformative practices include, in some form, the five elements of the worldview transformation model, as described in chapter 1. Any death-preparation practice can be transformative if our *intention* is to shift our fear of death into inspiration for living, and if the practice brings our *attention* to how we are currently living life. Through *repetition* of beneficial practices such as visualization and mindfulness, we can build new habits and new ways of engaging our thoughts and emotions. Making use of *guidance* from trusted authorities, established traditions, and noetic ways of knowing can help apply the practices to our actions and reactions to what life presents to us each day. In this way, we may foster both *acceptance* and appreciation for the ways in which death is a natural part of life.

#### ‹ PRACTICE ›
## Acceptance and Self-Compassion

Sit quietly and focus on your breath. As you do, contemplate death, either your own or someone else's. Maybe you picture yourself with a life-threatening illness, or perhaps you see yourself caring for someone who is at the end of life.

Breathe into your experience. Notice the feelings in your body. It's okay to feel fear. Bring an appreciation to the degree to which the little animal inside of you doesn't want to die.

Consider what happened to our ancestors who needed to survive in order to pass on their genes. The ones that passed on their genes struggled to stay alive. It's okay to have that part of you present. That part of all of us calls for compassion, for a sense of expanded awareness, and an internalized sense of goodwill and kindness for ourselves. As much as we seek to be kind to all beings, we should also aspire to be kind to ourselves. Just try to calm down and soothe those ancient circuits in the brain that are woven into the body and that don't want to die. Give yourself permission to feel scared and to not want to die or experience the death of another. Remember not to feel ashamed, as if you were a bad spiritual practitioner or didn't read the book correctly. Instead, sit with a feeling of acceptance and compassion, breathing it in and simply watching your own thoughts and sensations.

Feel the confidence that comes when you accept the deep wisdom that is within you. Continue to monitor your breath. When you're ready, take ten minutes to write in your journal about what you experienced.

# Chapter 7

# GRIEF AS A DOORWAY
# TO TRANSFORMATION

In order to really negotiate this path,
the first thing that a person has to do is
commit to doing the work. Then people
have to have humility enough to let go of
what has served them in the past and be
willing . . . to be open and innocent again
to what is coming. There's a lot of hard
work in transforming yourself.

LUISAH TEISH

One winter day in 1998, I sat in a large circle on the carpeted floor of a studio in Northern California. It was the office of Angeles Arrien, a transformation teacher who had donated the space. We were a diverse group that included religious and spiritual teachers from omnifarious traditions. There was a subgroup of scientists recording the event.

I'd assembled these people. I wanted to learn about consciousness transformation from them. I was examining the art and science of transformation in everyday life. I hoped to learn what triggers life-enhancing change, what sustains it, and how it impacts the ways we understand living and dying. I was humbled by all who showed up. Many were from the San Francisco Bay Area. Some curious colleagues traveled great distances to join this confederation of wisdom holders and spiritual seekers. As it turns out, even the teachers of life's transformations can be lonely. Being in community offers a source of vital healing.

We shared the stories that defined our lives. All of us were eager to find truth and insight from each other's lived experience. Even when our worldviews differed and conflicting views emerged, these bridge people sought common paths across great divides while sharing their own unique life stories.

When it was her turn, Luisah Teish explained that her calling as a teacher really began when she conceived a child. Birth was a portal into something profound, as all mothers know. For her, it was "primal and ancient and common to everything." In carrying this child, she was dedicated to a life she couldn't see, holding tightly to a deep and powerful mystery. She described the experience of laboring for twenty-three hours, working for something to be born. Twelve hours after the baby was born, she watched her beloved child die. "I often think: one hour short of a day to birth it, half a day of life, and then it dies."

In the process of living with the death of her child, Teish experienced a deep shift in her worldview. The shift was not easy. The catalyst was pain. Something in her died with the baby, she explained. But in the process, something new was born; she became a teacher and guide for others.

"In order to really negotiate this path," she reflected, "the first thing that a person has to do is commit to doing the work. Then people have to have humility enough to let go of what has served them in the past and be willing to listen to the message of spirit, to spend time out in nature, to let go of and be open and innocent again to what is coming. There's a lot of hard work in transforming yourself." She laughed.

## THE PRACTICE OF GRIEF

Like Teish, most of us have lost loved ones at some point in our life. One of my interns, a twenty-two-year-old undergraduate, said, "I guess I've been to more funerals than weddings." As our lives unfold, we lose more and more loved ones when the aging process takes its toll. Confronted with the death of someone we care for, we can be overtaken by myriad emotions that rock our world and our worldview. Our bodies, minds, and souls feel the loss. Certain events or times of year can remind us of someone we miss. And, depending on their worldview, some people report the feeling of excitement for what they perceive as a grand adventure for their departed loved one and ultimately for their own encounters beyond death.

The worldview transformation model emphasizes the role of transformative practices to help us live deeply and fully. These practices can take many forms, from a formal meditation practice to a mindful approach to gardening or walking in nature. As we have discussed, transformative practices include intention, attention, repetition, guidance, and acceptance. These five elements can also be applied to grief as a transformative practice.

First, we can bring our intention to learning and growing from our pain or loss. We can also shift our attention to our inner noetic experiences and how we may stay connected to our loved ones in our hearts and minds. By engaging in transformative practices in a systematic and repetitive fashion, we can build new habits or responses to our own fear and grief. We may discover the inherent resilience we have within us to overcome grief. We can learn to trust the hardwired capacities we have for survival—and for flourishing through our grief awareness. As Teish noted, it can take hard work to transform yourself, but in the end it is worth it. Ultimately, the gift of acceptance allows us to experience life on its own terms. Transformation is not so much about shifting the outer world, although that is part of the model. First, it's about changing the ways we respond to the complexities of life, just as it is.

## PHASES OF GRIEF

There are various models of the grief cycle. Elisabeth Kübler-Ross, in her classic book *On Death and Dying*, identified five stages: denial, anger, bargaining, depression, and acceptance.[1] This conventional model, now widely accepted, may not be the full story. In his 2009 book *The Other Side of Sadness*, psychologist George Bonanno notes that this model neglects our inborn capacity for resilience.[2] He argues that we don't graduate through static phases of grief. Instead, the majority of people are actually hardwired with a capacity to begin again. While grief is painful, it may also be short-lived. The process of grief may be both cathartic and transformative. According to Bonanno, it is like an oscillation, involving a range of emotions and ways of expressing our loss.

In an essay entitled "The Transformative Power of Grief," John Schneider offers his observations about grief and resilience.[3] The first phase of grief, he says, is discovering what's lost. How did we ever get by without having this person in our lives? How are we going to cope with them not being here? We hold on, Schneider observed. Or sometimes we let go in ways that are about escapism or lovely denial. We tend to define our grief by what we no longer have, rather than by what the departed loved one has given us.

The second phase of grief involves discovering what's left. Healing and growth may come when we allow ourselves to risk again, to continue growing, and to invite in transformation. The third phase involves discovering what's possible. Out of loss or separation comes a shift in our perspective or worldview. We can find a connection to something broader than the physical aspects of our being. As we move from "me" to "we," we may find a realm of love and interconnectedness. This process of developing other-regarding virtues can lead to both our personal understanding and a deepening of our relationships with others.

## FINDING THE WISDOM PATH

At the start of the journey, grief is painful. For Karen Wyatt, her personal pain became an opportunity for growth. Wyatt is a leader in the area of whole-person healthcare and a family medicine physician who has worked in hospices with dying patients for years.

> I found my way to hospice after the very tragic death
> of my father by suicide. I was still fairly young in
> medical practice at that time. But I really found myself

overwhelmed with grief and guilt because of his death. I wasn't able to recover from that, and ultimately it led me to start volunteering for hospice, thinking that maybe if I exposed myself to death and dying and grief and sadness, I would find my way through and out of my own grief . . .

In many ways I was numb, just going through the motions of life every day, even as a mother and a wife and a doctor. And I so desperately wanted to change. I wanted to grow. I wanted to heal the grief. I wanted to get to the place I saw those hospice patients in, a place of reverence and gratitude and appreciation for life.

So I set out to learn what it was that they understood about life that somehow I was missing. And I did manage to do that. I found a path for myself to transformation, a path that ultimately healed the grief and the pain that I was in, but also helped me wake up to life . . . and to appreciate the joys and beauties of life, as well as to manage the suffering and the pain of life. The path that I found I think of as the wisdom path. . . . Indeed, what I found was a life-changing experience.

Over time, Wyatt became a seeker of wisdom and insight about the nature of self, the meaning of existence, and the essence of living and dying. Her process of grief became a path to self-realization.

I was in a situation where I had plenty of knowledge, from all of my medical studies. I'd done lots of reading. I'd attended workshops and gone to counseling, and then yoga. I had plenty of knowledge to heal my grief, but I didn't have the wisdom that I needed at that time, and that's one of the reasons that I hadn't been able to heal. So working with dying patients who had faced their own mortality and were at the end of their lives gave me the wisdom I needed to heal my grief and change the way I lived in every moment of my life.

Wyatt's experience confirms the process that is described in the worldview transformation model. Her first-person, noetic experience of pain led to a time of opening. She was forced to reevaluate her life and what gave her meaning. She pursued a process of exploration and discovery. It was not until she began to work in earnest with her dying patients that she found a transformative practice that shifted her from the "me" of her own suffering to the "we" of connecting to a greater whole. She explores what she has learned in her book *What Really Matters: 7 Lessons for Living from the Stories of the Dying*.[4]

## WHILE DEATH IS INEVITABLE, HOW WE APPROACH IT IS NOT

It is important that we are able to transform our grief and see it through its various phases. In chapter 1, we met Margaret Rousser, who works at the Oakland Zoo in California. In our interview, she told me about Nikko, a male gibbon, who was showing signs of grief at the loss of his mate of over twenty-six years. He had a noticeable reduction in activity. For a time he stopped singing, an activity that is characteristic of gibbons in the wild, because his duet partner was gone. Rousser acknowledged that grief is a natural part of dealing with loss. But out in nature, grieving too long is a profound dis-advantage that threatens an animal's survival. This is why animals, from gibbons to humans, are hardwired to move beyond grief and loss, or in Rousser's words:

> It's important for animals to move forward just as it is important for us to move forward. Animals who grieve too long and have that loss of activity for too long are putting themselves at risk for predators. So I think in many ways animals are designed to move on. You know, life goes on. And you've got to keep going—you've got to keep living.

## PRACTICES FOR TRANSFORMING GRIEF

Over the years, I have experienced the loss of family, friends, and colleagues. It has not been easy. But over time, through my own personal practices and also through the connections to people who offer great inspiration and insight, I have learned to transform my own grief into a gift that has served to help my life and my work in the area of death awareness.

There are many grief practices that can facilitate worldview and personal transformation. The remainder of this chapter focuses on two general categories: noetic grief practices and shared grief practices.

## NOETIC GRIEF PRACTICES: ATTENDING TO OUR INNER EXPERIENCE

Noetic grief practices are those that include reflection, imagination, and spiritual insights that allow us to hold our loved ones close to our hearts while living our lives more fully. While there are many paths to transformation, we consider three useful practices in this section.

### Walking the Labyrinth

In her first book, *Walking a Sacred Path: Rediscovering the Labyrinth as a Spiritual Practice,* Lauren Artress helped catalyze what is now known as the Labyrinth Movement.[5] The growth of this movement, inspired by an ancient spiritual practice, is evidenced by the proliferation of these contemplative paths at medical centers, places of worship, and community centers worldwide.

The pattern of the labyrinth is usually about a forty-foot circle, Artress says. "It has one path that starts in the outer edge and weaves in a very circuitous way eventually into the center." Walking the labyrinth is a form of walking meditation that can lead us inward. As Artress explains, walking the labyrinth can be like visiting a watering hole for the soul.

> By walking, the mind quiets much easier for many of us. It's very much a Western tool. The labyrinth

is a path of prayer. I very much understand that
chaos is really just uncommitted energy. Prayer is
a practice that helps commit that energy through
intentions, through being in alignment with your own
integrity. All of that helps direct the energy toward
manifestation and prayer.

Artress also addressed some of the assumptions associated with grief.
Sometimes people feel a need to process grief along a fixed timeline.
"A lot of times people think, 'Oh, it's been a month, it's been two
months, I should be done,' instead of realizing that grief has its own
dynamic process," she said. She urges people who experience loss to
pay attention to their grief.

We're in a culture that thinks things need to happen
fast, and they should be over quickly and that, if we're
carrying pain, there's something wrong with us. And
so the grief process is really something to be honored.
I highly suggest bereavement groups, whether it's
through your hospital or church.

That's where a labyrinth can be helpful. Because
[when] you're walking, maybe you're crying while
you're walking; you're letting go, releasing, and then
all of a sudden, your tears dry up until the next time,
until the next round, until that well fills back up with
tears of grief. To have a place to release that grief is
really such an important touchstone for people. It's
really helpful with bereavement.

When you're walking a labyrinth, go in and find
your natural pace. Drop into your rhythm. That's
unusual, and the first time walking it might take a
little while. We really are forced one way or another,
quick and hurry and slow down and stop. Find your
natural pace. Then just draw that [departed] person to
mind or just start speaking to that person. Or use it as
a path of prayer and pray for that person.

Artress noted that walking the labyrinth is "not a belief-based process."

> It's simply releasing. When you're walking in the labyrinth, you're quieting and letting down. When you're ready, you're shedding thoughts. You may draw that person to mind. Just see where it takes you. . . .
>
> The universality of the human situation is often what gives people comfort. I think people who don't transform are people who are caught in a victim place. If you think your pain is more precious than someone else's pain, you're in trouble. And because of the human condition, we're all in this together. Our pain may be unique, but what happens with the labyrinth is that the broader picture is there.

## Laughter Yoga

Jennifer Mathews is a certified laughter yoga trainer. She has taught people to free up their laughter as a way to navigate daily life. She uses laughter as a transformative practice.

In 2011, her life partner, Kate Asch (also a laughter yoga teacher), was unexpectedly diagnosed with advanced cancer at the age of forty-one. She died only twelve weeks later. As the months after Kate's death went on, laughter became one of the tools Mathews used to move through the more difficult moments. She shared with me:

> After Kate died, I was faced with the true test of what we both taught—using laughter as a way to cope with challenges and shift energy. I remember one day when I was driving home, I felt the heaviness of missing Kate and knowing I wouldn't see her when I walked in the door. I thought, "Well, Jen, here's your chance to practice what you preach!"
>
> As I drove down Mount Shasta Boulevard, I decided to experiment. First I just told myself to smile. Then I faked a soft laugh. I certainly wasn't feeling happy in

that moment. But I decided I would laugh for at least ten seconds. Laughter yoga is a body-mind practice, and so I simply asked my body to laugh. All it takes is willingness. Before I knew it, the laughter grew and became more genuine. I could feel my whole mood change. I wasn't focused on the future I'd never have with Kate. Instead, I was enjoying the present moment.

Mathews told me that in laughter yoga, you laugh for no reason and "decide" to laugh whether you feel like it or not. When you do, biochemical shifts occur in the body and brain. She continued:

Some theories suggest that laughter brings us into the present moment because we shift out of the prefrontal cortex and into the limbic area of the brain, which is interested in the now. From my experience, we tend to feel grief mostly when our attention is on the past or future, on regrets or not seeing someone again. Since laughter brings us into the present and releases what we're holding on to, it can be a simple but powerful tool for healing. Plus, in my opinion, giving ourselves permission to enjoy life is one of the best ways to honor those who have died.

Mathews's current lifework is sharing what has most supported her own grieving process with those who are facing the loss of a loved one. She believes death can be the ultimate inspiration for true happiness.

Death dares us to find internal joy despite the external circumstances of losing a loved one. Choosing to laugh is a way to foster that and to break the pattern of following thoughts that lead to sadness or grief.

## Communion with the Dead

Communing with the spirits is a practice that is used worldwide to transform the fear of death and the pain of grief. Reflection, prayer, and shifting our awareness to our voice are tools that people in many cultures use to connect with unseen realms beyond the living. This practice can help shift grief into an ongoing sense of connection with the departed.

### Consulting Ancestral Spirits

Drawing on her own Yoruba Lucumi tradition, Luisah Teish explained her method of engagement with spirit, which she calls upon daily.

> Once you have gone through establishing a relationship with the surviving intelligence of your ancestors, the feeling of being alone is gone. . . . Ancestor reverence for us is a very practical thing. It's not only a recognition that I've got my body, my personality, my knowledge, my property . . . from the ancestors. That's a given.
>
> But the reality for us is that every thought, every emotion, every experience that every human being has ever had still exists as capsules of energy that I can connect to through the proper rituals. . . . I can say, "It's time for me to prepare dinner. I now invoke the intelligence of all you garden cooks and waitresses to come help me get this done." Once a person knows how to do that, a sense of security, even in the face of disruption, is there with you.
>
> The biggest one for me is overcoming indecision: "Shall I do this, shall I do that, shall I do the other?" Boom, you talk to the ancestors about it, and oftentimes you will get a perspective or some information or a feeling that you would not have had if you had not connected to that. So there is a point where you learn a simple system of invoking and divining.

### Spiritualism

In the West, another practice for communing with the departed is through mediumship. People in grief may seek to communicate with their departed loved ones through an intermediary. Julie Beischel, a scientist who works with mediums, observed that consulting spirit guides can be helpful for healing and processing grief (see chapter 5). As we learned in chapter 3, many people report being visited by the spirits of departed loved ones. But those people who don't experience such communications spontaneously can be prescribed a mediumship reading, according to Beischel. From her experience, she believes a combination of psychotherapy and spirit communication may work best for addressing grief.

Beischel explains that one way you get through grief is by redefining your relationship with the person that you've lost. Working with a medium to communicate with someone who has passed on can show you that your loved one is still in your life, just in a different way. This understanding may be very comforting. You can then bring this new understanding to a mental health professional, who can help you with the process of redefining your relationship with a loved one who has passed.

Beischel told me:

> We hear individual stories that are very compelling.
> We worked with a sitter . . . who had lost an infant
> son to congenital heart problems. Decades later,
> he had another grown son who died. He said after
> the devastating loss of two sons . . . his life was
> unbearable. Then, having received messages from
> his children through mediums, his life is not only
> bearable but worth living. That would sell big if
> you could put that in a pill. The patent on that
> would be very valuable. I believe it's a prescribable
> treatment option.

## SHARED GRIEF PRACTICES: FOSTERING COMMUNION AND CONNECTION

Shared grief practices are those that invite us to reach out and connect to others through communion, celebration, and collaboration.

Knowing we are not alone allows us to build a web of interconnections that make us strong and resilient. We move our concept of death from the "me" to the "we" and, in turn, join fully in the transformative nature of life.

All cultures have ways of honoring the dead. Funerals and burial practices are important elements in the expression and transformation of grief. Sharing our loss through such practices is a powerful way to shift our relationship with death. Such practices can take many forms.

## Remembering Together

Returning to the place we were born and raised can be an emotional experience. It was for me, because of family members who are now passed on. I took my son, Skyler, on pilgrimage from our home in California to my home of origin in Detroit, to pay final respects to my mom and stepdad, who died two years earlier. My sisters had arranged a memorial and burial service. It was a sweet good-bye for these two beautiful people who played such an important role in our lives.

The minister shared his words of kindness as he guided the brief memorial: "And I am watching the people that I love leave, and I am being left behind by the people that I love. And then I have to say the hardest sentence there is to say. I have to say, 'Good-bye, I love you.' How can it possibly be, good-bye, I love you, in one sentence, together?"

My stepfather fought in World War II. At his burial, he was entitled to a twelve-gun salute. Dressed in their military attire and armed with guns, the soldiers gathered outside the small chapel and fired their powerful statement to one of their own, now fallen from this world. Following this powerful ritual, we made our way to the family gravesite, where we would bury their ashes. Standing in the cemetery with all of my predecessors, my ancestors, I thought about this mystery of life and of death. While we try to avoid discussions about death in our culture, when we come to a cemetery and feel the connection to all the people who went before, death connects us. Life does go on within us and without us.

## The Wake

Luisah Teish told me how the dead were honored in her Yoruba Lucumi tradition:

> Throughout the African diaspora, there are very elaborate, threefold rituals to honor death. They address care of the corpse, care of the spirit, and care of the community that's left behind.
>
> The wake involves sitting with the body. This is a time when everyone in the community is going to bring food, the children are all going to be there, and people are going to sit up all night having a wake for this person. Then we had a memorial where we reviewed that person's life. All of these things are for the community that's mourning.
>
> I do what's called a nine-day elevation to help that spirit leave here and know that they are somewhere else. We have songs that we sing for that. We have things that we cook and rituals that we do. It's all very elaborate.
>
> One of the really important differences that I've learned over the years through teaching in multicultural situations is, although everybody addresses the corpse, the spirit, and the community, what kind of emotional attitude there is differs from culture to culture.

## Memorial Day

In 2012, I visited the Cypress Hill Memorial Park in Petaluma, California, for the Memorial Day observances. There, I talked to many war veterans, primarily from the Vietnam War. They carried their scars and hidden wounds. From them, I heard stories of loss, guilt, ghosts, and their reasons for remembering their fallen comrades on this emotional day.

Keeping people alive in our memories is a powerful form of expression. This was confirmed by one Vietnam veteran who participated in the Memorial Day event:

This is for the ones that gave all. That's what this whole day is about. It's a remembrance and honoring. I think the biggest thing is just to remember. We all served, but we are the ones that got to come home. And a lot of people that are buried here didn't come home.

I think it's very, very important that we all remember the people who gave the ultimate sacrifice going back to all the wars—World War I, World War II, Korea, Vietnam, Desert Storm, Afghanistan, Iraq. People are still dying today, and we need to let them know that we're not going to forget.

With cameras rolling for the recording of our film, *Death Makes Life Possible,* another veteran told me what it means to him to share in our collective loss and how that loss raises personal questions.

Today is a remembrance. You think about the big question. Why did we come back and the others didn't? I was downed twice. You wonder why you make it and maybe the pilot didn't.

Fear around death is different for every person. Some may not even think about fear when it comes down to grabbing your weapon and defending yourself. You do it to survive. Many times you'd see your friends get hit that are sitting beside you. I've had several pilots come out from under me and the fear hits afterward. That's when you feel it.

Breese Baker was trained as a medic when she joined the military in 1966 and headed for Vietnam. Part of her training involved maintaining an objective stance toward her patients as they faced death. Having emotions about death is not part of military indoctrination, she recalled.

It's a job, and that's all they want you to do. But what happens is that, when you have people, especially the

men who are coming in off the field and have been
mortally wounded and not expected to live, your own
personal feelings start coming in, and you start caring.
There isn't any other way. For me, I just had to stop
and just know that in that moment I had to do this. It
really brings a tear to your eye.

Doing this and coming here, I just have a sense
that there is someone that lingers here. I've always had
the feeling that I can walk into a place and know that
there is somebody there. It's like the hairs come up,
and you just kind of get that feeling like, yeah, there is
somebody here.

Many of the other veterans expressed an ongoing connection to
those who died in battle during their tenure in Vietnam. Being part
of a group is very important for military veterans. One veteran told
me that even in death, the departed soldiers are aware of the honoring that was taking place amid the grave markers.

There's a special camaraderie with veterans. When
you're in the service, they're your closest people.
When you rely on somebody with your life, and
they're relying on you for their life, there's no closer
bond that you can have. You may never ever see them
again after your service is over, but it always stays.

Even after death I don't think you lose that
camaraderie. You're always a part of that fraternity
of being a person in the military. You never lose
it. I've been out of the military for almost forty
years, and you still never lose that. You just don't.
It becomes part of you.

That sense of belonging may be even more
important after death. I think all the veterans here
know. They've all known about Memorial Day and
Veterans Day before they've gone, and they know that
it means so much to remember. So I think they know.

I think they know that we're thinking about them, and that every year, they know that those American flags are on their graves.

They're proud in their grave. They gave their life for their country, and they're proud of it. They will always be proud of what they've done. I think it goes on through eternity.

## Día de los Muertos

One of my favorite celebrations occurs every year on November 1. In the weeks before this date, the community where I live embraces the Mexican holiday of Día de los Muertos. While originating in Mexico, this celebration of death draws people from diverse ethnic and cultural backgrounds. It has origins in the indigenous communities of the pre-Columbian past.

On the holiday, friends and family gather to remember their departed loved ones. Those who participate in the tradition believe that the veil between the living and the dead becomes thin, and the dead become more active during this time when fall is upon us and the harvest season is complete. In Petaluma, California, as in other cities in America, Mexico, and throughout Latin America, people build altars to honor their departed loved ones. Offerings, including drinks or food, marigolds, and candles, are placed on shrines. These are items that the departed spirits will enjoy. Altars also typically include pictures to help people remember those who have passed. As one celebrant explained to me, creating the altar "gives us an energetic connection to our loved ones."

Día de los Muertos rituals include dynamic parades and boisterous celebration. There is a palpable sense of joy and conviviality. People are dancing, playing music, singing, and talking to their departed ancestors. It is a way to both honor the dead and appease them, so that they don't cause trouble for those who remain in the realm of the living.

Several months after my mother's passing, I found myself in downtown Petaluma on November 1. I was drawn to the colorful costumes and macabre images of skeletons and caskets. Women were dressed as the popular icon the Lady in White, also known as Catrina, a female skeleton dressed only in a fancy hat. Children stood inside giant puppets of *el diablo,* a reminder of the dark forces that accompany death. A mariachi band played near the bridge crossing the Petaluma River. A little farther, a band of marching skeletons assumed formation.

My friend Gloria MacAllister (her last name from her Scottish husband) greeted me with enthusiasm. She is a cultural catalyst who bridges her Mexican heritage with her California life. With Gloria, I was caught up in the excitement, marching with the flow of spirit and inspiration. I carried a candle to honor my mother. Gloria was dressed as a spirit. To my left was Catrina. She was a haunting image that pervaded this traditional event. The puppets changed hands, and I helped Gloria carry one of the large, heavy papier-mâché constructions. The weight was noticeable; honoring the dead is not a light and easy process. And the load was shared—for all are responsible for the ancestors. I felt a heaviness in my heart at my mother's recent passing. This soon gave way as the parade continued and I found myself going with the flow. Death does make life possible.

## GLEANINGS

As we have seen, people engage in different practices to honor their lost loved ones. Such practices are used to help transform grief into an honoring of life. In this process, grief becomes a tool to help us grow and thrive in the face of death. We have considered inwardly directed, noetic grief practices that help us find our own path to healing. We have also considered shared grief practices that involve communion, celebration, and collaboration. Sharing grief in the form of a collective practice moves us out of our small identity into a larger meaning system. It helps us see that we are not alone and that we need not forget those that have passed, for they are alive in our hearts.

These tools help us in our own individual transformation, giving us insight and comfort. We may use them to redefine our relationships to those who have died. We may also use them to connect with our broader community to celebrate both the dead and those who are here to remember them.

Loss is a complex issue. It can force us out of our steady state. In doing so, it can shake our worldview at its core, disrupting the parts of our life that we've taken for granted. We may deny the pain. To be sure, denial is a natural and highly valuable defense mechanism. Yet grieving is a natural process that takes time; it has its own cycles and phases. And as the worldview transformation model predicts, transformative breakthroughs can emerge from our suffering. Profound loss and grief, held with intention and attention, may create a dynamic caldron in which to birth new understandings of who we are and how we relate to others. As the famous poem by Rumi tells us, "You're not just a drop in the ocean. You are the mighty ocean in the drop."

◂ PRACTICE ▸

## Finger Labyrinth

Find a comfortable spot where you can put down this book and make use of the finger labyrinth shown in figure 2.

Begin by taking three deep breaths. Enter the labyrinth by placing your finger at the opening. Trace the pattern with your finger, bringing attention and intention to your actions.

As you do so, contemplate the loss of someone you have loved. Feel your ongoing connection to them as you continue the finger movement. You may wish to talk with them in your mind or send them loving intention. Or simply hold your loved one in your heart while you clear your mind and attend to the pattern under your finger. Notice how you respond when the labyrinth offers challenges or makes you feel that you've lost ground on the path.

When you reach the center of the labyrinth, pause to reflect on what you are feeling in your body. What kind of sensations come to you? What thoughts and emotions arise for you?

As you retrace your journey in reverse, winding your way out of the circular design, release any tensions, fears, or sadness by taking deep breaths as you complete the practice.

When you have completed this practice, take ten minutes to express what came up for you by writing in your journal. In particular, ask yourself how grief can transform you in ways that allow you to experience gratitude for what you have learned.

**Figure 2**  Finger Labyrinth

# Chapter 8

# DREAMING AND THE TRANSFORMATION OF DEATH

When we finally step out of our bodies and
become like a consistent, constant dream,
hopefully we discover our hand on the rudder
as we're moving through those states.

LAUREN ARTRESS

Dreams are a powerful tool for transforming our relationship to death. They can be used to process our grief, to enhance our feelings about or relationship with our departed loved ones, and potentially, to prepare ourselves for the great adventure that lies ahead.

As I've been immersed in the writing of this book and the making of the film *Death Makes Life Possible,* my dream life has been active. Some dreams are hopeful and filled with optimism; others speak to my own fears and anxieties. In one dream, I arrived home after a difficult journey. The house was some version of my childhood home. Much to my surprise, all the lights were on, and the doors were wide open. I was afraid because my mother was there and should have been asleep by now. As I drove up the driveway, I found her sitting in a small porch area across the driveway from the main house. She was chatting with a colleague of mine. He had helped with the filming of *Death Makes Life Possible,* and during those times, we had shared our worldviews about death. In the dream, my mother and my friend sat together as cozy neighbors. The tenor of the dream was total peace and goodwill. I was struck by how well they got along and how comfortable José was in my mother's presence, even though she had died several years ago. I greeted them as I pulled up the driveway. My mother pulled me aside and told me how she enjoyed José's company. We hugged. I then continued on to the dream version of our house at the back of the property. My best friend from high school was there. She was apparently my roommate in the dream. I shared with her the way in which my mother's youthful charm was palpable and her connection with José so clear, even though she was dead.

When I awoke, it took me a few sleepy-eyed moments to recall that my mother is no longer alive. I was nonplussed. She had been fully animate and vital in the dream. I also wondered about José, recalling all the dynamic interviews and conversations we've shared over the years. He is a man with great depth and an enlightened stance toward death. My heart was filled with delight and gratitude that he helped my mom.

This dream has stayed with me, providing both a link to my now-deceased mother and an appreciation for my friend. As my friend Luisah Teish explained, "Dreams are the places where the ancestors speak to us. Dreams are the places where it's clear that the dead are not dead—that they're connected."

## DREAMS AND OUR SHARED HUMAN EXPERIENCE

Dreams are mysterious. They offer glimpses into our inner worlds, using symbolic language that may be elusive and veiled. They may be difficult to comprehend even as they reflect our direct inner experiences and our cultural beliefs and shared worldviews. We often look to dreams for their personal meaning. The representational nature of dreams, grounded in rich and multidimensional metaphor, can be a window into our individual experiences and responses to the physical world in which we're embedded. They may also connect us to the immaterial nature of life and death.

Throughout history and in every reported culture, humans have given great attention to their dreams and their meanings. Many cultures highly regard dreams, and individuals use them to inform their daily practices and long-term plans. In early biblical writings, dreams were accepted as communications from God or the devil. The sleep temples of ancient Greece used dreams to treat sickness and to help diagnose diseases. William James respected the phenomenal nature of the dream as a form of existence unto itself. He wrote in his beautiful prose:

> The world of dreams is our real world whilst we are sleeping, because our attention then lapses from the sensible world. Conversely when we wake the attention usually lapses from the dream world and that becomes unreal. But if a dream haunts us and compels our attention during the day it is very apt to remain figuring in our consciousness as a sort of sub-universe alongside of the waking world. Most people have probably had dreams which it is hard to imagine not to have been glimpses into an actually existing region of being.[1]

In the context of death, dreams have many functions. Dream imagery may help us to integrate major life changes, including the death of a loved one, our own impending death, or our general death terror. They may help us feel a connection to those we have lost. Dreams are tools for understanding what we need for our own growth and development—offering a gateway to personal transformation. Emotions and physical sensations are expressed in dreams, and these feelings can offer insights into our own relationship to death and what may happen beyond. They may offer a means of integrating our experiences with death into a new worldview that bridges life and death. Finally, they may be a means of communicating directly with the departed.

## EMBODIED DREAMS

I met with Kathy Chang-Lipsenthal about two years after the death of her husband, Lee. Kathy wanted me to help her make sense of a couple of dreams she had had.

The first dream was simple, but deeply nourishing. She felt her husband in the bed with her. He held her, and she felt a renewed sense of connection. In the morning, when she awoke, she felt happy.

"Was that really him?" she asked me.

"Whatever it was, it sounds beautiful," I replied.

In her second dream, Kathy was in a large room. The ceiling went up and up. There was a pool in the center of the room. Suddenly, Lee was there with her, and he took her up to the top of the room. He knew she was afraid of heights, but he insisted. A small path served as a kind of bridge from one side of the room to the next. Kathy was scared. Lee surprised her by pushing her out into the open space. As she fell, he came from behind her and held her, guiding her to fly from one side to the other. She hoped that they would land in the water, and she was relieved when they were over the pool. But Lee kept guiding her between the comfort of a smooth landing in the pool and the fear of a painful landing on the hard concrete.

This dream made Kathy feel that Lee was still with her, pushing her to overcome her own fears and to live to the fullest. She also felt that it symbolized Lee's continued support and guidance. The dream helped her to recover some of their intimacy in the course of her own growth and transformation. By sharing her dream experiences with me, Kathy invited me to collaborate on a meaning-making process that served both of us by connecting us to Lee.

## TENDING OUR DREAMS

Stephen Aizenstat, PhD, is a psychologist and the founding president of Pacifica Graduate Institute. The focus of his work is depth psychology and what he calls "the tending of the soul." Depth psychology is rooted in the work of Carl Jung, William James, and Sigmund Freud and focuses on the relationship between our conscious and unconscious mind. This area of psychology rests in the domain of the psyche, that ineffable quality described as soul, mind, or spirit. Aizenstat has developed a specific practice for working with dreams: the Dream Tending practice. In it, he encourages us to work with dream content as "living images." He explains:

The wisdom of ancestral callings, the instinctual
knowledge of animal visitations, the musings of the
soul are attended to from a psyche-centered, rather than
person-centered, perspective. The "intelligence" of the
dream is listened to from the inside out, accessing the
innate knowledge indigenous to the figures of imagination.

Dream Tending, according to Aizenstat, moves beyond the more causal and reductive modes of analyzing or interpreting dreams. Instead, dreamers experience the content of dreams as dynamic and alive. Dreamers are instructed to look deeply into the messages from dream figures, including images of landscapes and objects, in order to gain insights and perspectives. Aizenstat's practice moves beyond examining human communication in the dream world to connecting the dreamer to nature and the unseen realms of psyche. We are all a part of "nature's dreaming," says Aizenstat, and when we experience the psyche as part of that, we experience the life-death transit as integral to particular dreams and to the dreaming process in general. In Kathy's case, for example, she was able to find meaning in her dream encounter with Lee. Using Dream Tending can offer both intimacy and empowerment.

Using a method of association, each of us can tend to our dreams in order to open ourselves up to the symbolic nature of our own personal experiences. Exploring our dreams can offer new insights into our waking life. In the process of amplification, we as dream tenders can examine the content of our dreams to identify symbolic or archetypical aspects, in order to apply them to our understanding of ourselves and others. Using the method of animation, we may tend to the content of our dream experience in terms of our relationship to the world and our place in the earth's ecological systems. Each method involves what Aizenstat refers to as an "attitude for how we work with the dream."

He says that dreams are alive and filled with their own intentionality or purpose.

The idea is to tend the dream as if it were happening
now. If we are sharing a dream with someone, the

dreamer sits and tells the dream in the present tense. In retelling it, the dream comes into our present experience. Once the dream is told as story, happening in the here and now, dream figures come to life, become animated, and open themselves to a relationship between the dreamer and the dream figures. In this way of tending to dreams, images of death or dying become available to us for conversation.

There is a multidimensional nature to the dream experience, according to Aizenstat. Death may be speaking to us in metaphorical terms, rather than as a literal communication.

Death will often pick up that sense of ending. It takes us to a particular place of consciousness, like a descent into the realm of an underworld experience. In the place of Hades, the Greek god of the underworld, it is imagined that death is contiguous with all of life, informing life in every regard. Rather than being frightened of death, it represents something that we've been in contact with all through our years, sometimes consciously, most times just out of awareness. . . .

Our death is always part of us. We are programmed from the beginning to die. A genetic intelligence is always at work in the human condition. The dreaming psyche will pick up on that intelligence. The reaction or response in our culture is somehow to withdraw or be frightened. Rather than moving away or being frightened, I suggest that we really allow the image to be present so we can move out of our fear and into our relationship with death . . . as it opens us to the abundance of life.

Aizenstat's view is consistent with the terror management theory literature. He advocates that we bring our relationship to death into our awareness, even as it takes form in our dream states. He explained to me:

Death too, as a living image, brings its value into the natural fabric of our dreamtime. When repressed or denied, death as image pushes with ever more intensity and expression. When viewed as part of our natural interior landscape, death, as a companion of life, offers perspective and generative possibility. When the fire moves through a forest, room is made for new growth. As the seasons change, springtime brings the germination of waiting seeds and the blossoming of flowers.

Death as a companion, an active member of the dreamtime, brings us more fully into the abundance and beauty of our lived experience. We appreciate being alive—we stop and smell the flowers, cherish precious moments, and are less fearful of the inevitable and irreversible. In dreams as well as in awake life, when death is experienced metaphorically and companioned as a familiar in an ongoing way, we greet the world with renewed appreciation. When literal death approaches, we are less afraid, less horrified, and, in fact, we become better able to sustain our quality of life. We live in ways that actually draw on the resources and intelligence that death as image offers. We live longer and more fully into our dying with death as a welcome guest.

## LIVING, DYING, AND DREAMING

Fariba Bogzaran is a visionary artist and consciousness researcher. Her focus is on dreams, science, and consciousness. Her own dream experiences are fundamental to her life and work. She used the power of dreams to help her friend and mentor, Gordon Onslow Ford, during his dying process.

I had learned the auditory system works very well for people who are dying. I would talk to Gordon in his ear. Every time I would see his face stressed, I would say, "Gordon, relax. There's nothing to worry about. Whatever you see is a dream." So he would relax for the next couple of hours.

And then, he would just have a little more anxiety—at that moment people are probably having dreams and they are struggling, they don't know if it's an actual reality or if it's a dream.

I was just telling him in his ear that this is a dream and that he could relax into it. That was the moment that he passed. It was very, very beautiful. He loved walking in the woods, and the way his breath went slowly, as if he was walking in the woods and slowly vanishing. Just going—disappearing.

Bogzaran is also both an observer of dreams and a lucid dreamer. Lucid dreams have been reported to be powerful tools for death preparation and spiritual transformation. A lucid dream is any dream in which we are aware that we are dreaming. Bogzaran noted:

One of the major works around lucid dreaming is about preparing ourselves for dying. The core of that practice is about looking at our habitual behavior, whether in waking, whether in dreaming. What are the places that we're locked in and we're repeating the same habits? How do we unlock ourselves?

As the worldview transformation model predicts, building new habits is essential to a transformative practice. Like Dream Tending, lucid dreaming can offer insights for bringing mindfulness to all of life. As Bogzaran explained, in her worldview, the practice of lucid dreaming is about lucid living.

It is about becoming aware in waking—becoming aware of every minute of our waking. Every breath. Because what happens is that the minute you're talking about dying, you're talking about breath, one last breath. We come through this world with an inhale; we go out with an exhale. And there is a unique "in-between" that is suspended in the midst of the inhale and the exhale.

Lucid dreaming can serve as a path toward getting in touch with what Bogzaran called "that expansive aspect of one's self." She talked about synchronicities and meaningful events that reveal something interconnected about our existence. Lucid dreaming, like meditation and other transformative practices, helps people develop a reflective consciousness. In this state, we can witness, but not become attached to, the fears or anxieties that may be associated with death. As Bogzaran noted, "It takes time to really learn to observe, learn to witness." She said:

> Death is a taboo to talk about, and we sometimes think it's something that happens to somebody else. But death doesn't know culture. Death doesn't know religion. Death doesn't discriminate. That's the thing—it's going to happen to all of us, and we have to be prepared for it in any time, in any moment. For me, this is the moment. I'm here and I'm talking. This is the moment. The next moment I might not be here. So this is the moment when I have to be present. The next moment, I don't know. I might not breathe. I might just drop dead. I don't know.
>
> And it's the same thing with all of us. My practice with lucid dreaming is that every morning when I wake up, I tell myself, "Good morning, Fariba. You're awake. You're in this life. This is new." I'm not taking it for granted that I'm waking up. Every morning the practice is, I open my eyes to life. I have this one day. This is the moment.
>
> Every night when I fall asleep my practice is, "This might have been the last day, and I'm transitioning.

I'm just going to go into my dreaming consciously."
Just bringing that kind of consciousness to life is the
practice of lucid dreaming and lucid waking. . . . Of
course, I'd love to believe that there is something
greater than us on the other side, and [there] probably
is. If I make it into a theory, that wouldn't work.

## ARE DREAMS PREPARING US FOR DEATH?

David Hufford, a cultural anthropologist and author of *The Terror
That Comes in the Night,* is an expert on dreams and experiences of
otherworldly presences in dreams. He doubts that dreams are the
preparation ground for death. He explained:

I have always had a problem with the afterlife/dream
concept. My dreams range from nonsense (I do not
go for the Freudian view that replies, "Aha! Now we
are getting somewhere—tell me more about this
nonsense") to plotted vignettes to rare metaphorical
precognitive glimpses. Perhaps lucid dreams relate
through their apparent connection to OBEs [out-of-
body experiences], but even those are not very stable.
It seems that dreams include a lot of random nonsense,
perhaps brain (rather than mind) supplied, and the
comparison comes from the very diverse view of the
afterlife we get from reports, whether [from] mediums,
NDEs [near-death experiences], whatever. But that
diversity of reports is to a large extent attributable to
"noise in the signal."
Our accounts of the afterlife seem definitely to
include some stable observations of reality, but that
includes an imaginal (not imaginary) element that
complicates interpretation. And beyond that, there is
wish fulfillment, secondhand reporter editing, lies, et
cetera. I worry that the dream analogy is a slippery
slope. Not that I consider dreams unimportant on the

topic, but I do doubt that we have reason to think the afterlife operates in the way that dreams do. And using the dream image even as a metaphor for the afterlife simply aids the materialist skeptic's view, which makes all visionary experience into dreams vaguely defined.

Just as we have seen different views on death and the afterlife, there are various, sometimes conflicting views on the role of dreams in death. Our perspective—perhaps our experiences in dreams—are informed by our worldviews.

## DO BELIEFS SHAPE REALITY?

Biologist and author Rupert Sheldrake, like many of the religious scholars we have considered, believes that dreams can be a rehearsal for death, and argues that beliefs influence our dream state. If people's experiences are shaped by their lived worldview, so too then are the ways in which they can prepare for their dream life and eventual death. Sheldrake told me:

> I'm not sure what happens when one dies, but I think what happens largely depends on what one believes will happen. I think . . . when we die we go into a state which is like a dream state. The only difference . . . is that we can't wake up because we haven't got a physical body to wake up in. So if you're in a kind of dream state after death, then in a dream state you have a body.
>
> Each one of us practices every night for dying, in the sense that we have a dream body. There is another body besides our physical body, which exists in our dreams. We're in it several times a night, even if we don't remember. And it's not a regular, physical body, but it's modeled on our physical body. What happens after we're dead depends on our memories, our desires, our beliefs.

The question becomes, what do we believe about death? And even if ours is a metaphysical view that cannot be scientifically measured, what does it mean to how we live, grow, and transform? How do our dreams inform our views on death and our way of preparing for the inevitable transformation?

## LINKAGES TO THE DEPARTED

Dreams may serve as points of connection with our departed loved ones. They may be a message from a wiser part of the dreamer's psyche. That is the standard psychological view. It is also possible that our departed loved ones come to us in our dreams to help guide and instruct us.

As a member of the clergy, Lauren Artress, whom we first met in chapter 4, has been called to assist the dying and their families on many occasions. This has allowed her to formulate her own perspectives on dying and the role that dreams play in our connections to the departed.

> As a pastor, I've often experienced that, when a person's dying, they have what I would call bleed-throughs. They become connected or, you know, visited by a deceased relative who's there to be with them, perhaps guide them. This bleed-through into the other world as a beginning of the transition is really important.
>
> I think these bleed-throughs become part of the person's dreams. If you're dying you may hear, "Oh, I just saw my husband." He had died a few years ago, and the person may not be conscious of it. My mother, who's very elderly now, almost one hundred, has severe vascular dementia. She often feels visited by my deceased brother-in-law, who just died recently. They were living together before she was moved to assisted living and before he died. She feels like she's being visited. And so we go with that. "How is he? Glad he's visiting you."

I think after death, it could be a continual dream
state. Can we ever really control our dreams? I don't
believe so, because our dreams really come from
the unconscious. When we pass over, soul is in the
unconscious, or part of soul is unconscious. So that's
one of the questions I have. But when we finally
step out of our bodies and become like a consistent,
constant dream, hopefully we discover our hand on
the rudder as we're moving through those states.

Luisah Teish shared her worldview on the value of dreams for transforming our relationship to death.

Dreams are the places where the ancestors speak to us.
Dreams are the places where it's clear that the dead are
not dead. They're connected. . . . I think that we have
to appreciate that what we think of as invisible, or
subtle, is actually energy moving in a nonmaterial way.

Likewise, Tony Redhouse sees dreams as a transformative practice
for connecting with those people or spirits who have passed on. The
Native American healer believes that when we are dreaming, we are
in a special state of consciousness that makes it easier to connect to
unseen realms.

When we ask for a dream, the spirit is able to speak
to us because we're in a vulnerable state. When our
consciousness is shut down, we're not using our
rational, commonsense, practical thinking. That's why
the spirit can speak to us in that state. Sometimes, I
have so many things going on, my mind can be racing
a thousand miles an hour, and the spirit has difficulty
getting my attention. When I'm asleep, I know that
the spirit has a wide-open ability, an opportunity just
to drop a bunch of wisdom, to speak with me directly,
and to get my attention. I'm open to that opportunity.

For me, dying is simply coming to the truth of the matter. Everything that we've done our whole life, every relationship we've had, everything that we've hoped for, everything that we've dreamed is going to become crystal clear in that moment.

## GLEANINGS

Dreams are a mysterious way of accessing our unconscious mind and a way of integrating elements of our daily experience. They can offer us ways of confronting our fears and anxieties about death and of feeling connected to those who have passed on. The process of dreaming may help us to experience life beyond our embodiment and to encounter realms of noetic insights that broaden our worldview.

Diverse cultures approach dreams as ways of engaging our souls or psyches, because dreams are not part of waking experience and transcend our rational intellect. As with other transformative practices, we may engage our dreams with intention, attention, repetition, guidance, and acceptance, shifting our self-understanding and our relationships with those we have lost. Through Dream Tending, we may gain new insights into the hidden language of dreams. Likewise, in lucid dreaming, we can gain insights about the extended reaches of mind and consciousness that lie beyond the physical body. In this process of dreaming, we may begin to understand and direct our own concerns and hopes in ways that grant us comfort and peace. We may find ways to connect with those who have passed over; finding intimacy in our dream relationships can offer us sustenance and hope in our waking lives.

### ◂ PRACTICE ▸
### Dream Recall

Dream recall offers important insights into your unconscious mind. Recalling dreams is something that can be learned. Begin by noticing the amount of sleep you typically get. The longer you sleep, the easier it is to recall your dreams.

To help in dream recall, when you wake up during the night, ask yourself what you were dreaming about. Keep your eyes closed and don't move. Just invite your mind to answer. Be patient. Maybe what you recall is just a dream fragment. Start free-associating with that fragment, looking for details that may help you in your dream recall. Have your journal near your bedside and take notes about your dream experience. Observe the elements of your dream as though each component had a voice; this technique can help you to understand its hidden message. In the morning, review your journal and record your observations. Allow your dreams to speak to you and to aid you in your personal transformation.

# Chapter 9

# TRANSFORMATIVE ART

We are filled with grace and wisps of stars.
Do you remember who you are?

**GARY MALKIN**

A rt is a powerful tool for expressing our emotions and perspectives about the world. Through art, we can communicate our attitudes toward death and share lessons we have learned about how to lead a rich and meaningful life. Creating art can be a transformative practice that allows us to move through our own grief and suffering and to help transform the grief and suffering of others. Through various mediums, such as music, dance, visual arts, storytelling, and poetry, people from all cultures have found ways to explore their own relationship to death and the afterlife.

We need not be a talented artist to discover how art can bring us closer to our most authentic self. There is an artist in us all. In moments of deep transformation, art can become a faithful friend, there to help us express our emotions. The heart wants what it wants. The soul craves new sources of insight and inspiration. The loss of a loved one, grief, and rebirth are all part of a process that can take art to new forms. Art can transport us into an altered, expanded state of consciousness. It can offer us a new lens for perceiving who we are.

Art can serve as a bridge between our Newtonian world, based on cause and effect, and theoretical realms of imagination and entanglement. In the process, we may find the delicate balance between physical form and dynamic consciousness untethered to our material nature. Art can also be used to help create ceremonies and expressions of appreciation for living with dying.

Both developed artists and those for whom art is a foreign concept can find that art is a way to engage the mystery of life and death. Creating art may be a tool for transforming grief or a medium for celebrating the gift of life. It offers a creative mode of inquiry into consciousness and the essence of who we are, including the many ways we relate to death. When we bring intention to art as a transformative practice, creating art can be a means of embarking upon new habits or ways of being. We may find guidance in our creative mind. Ultimately, the practice of creation and artful living may help transform our worldview of death.

## LIFE AS CEREMONY

For Tony Redhouse, life is a ceremony filled with art and music. In each breath, this ceremony takes us beyond place, time, or events. As a recording artist and performer, Redhouse intuitively sees a connection between music and the healing rhythm of life. He says that art connects us to the spirit of life—and death.

> All indigenous people around the world have used
> the voice, the drum, and the flute to take them into
> ceremony, to be able to express what is the deepest
> part of their soul. I have taken those same instruments,
> and I have used those to be able to connect people
> with a spiritual space, with a sense of spirit. . . .
>
> When we hear the ancient sounds—when we hear
> the heartbeat, the sound of the flute, when we use
> the resonance of our voice—it takes us back to our
> beginning. It takes us back to what we were before
> all the technology, before all the books were written,
> before all the governments, before all the institutions,
> before language, before all of the different eras that
> we have gone through in this universe. We go back to
> that. We go back to that beginning place. And that's

what all of us are wanting to do. We're wanting to go back to that essence, to the truth of who we really are.

Each day is a prayer. And every action that I do, every thought that I think, every dream that I have, is creating that ceremony for me. What kind of ceremony do I want? What do I really want as a ceremony? I'm seeing that happening and becoming a reality with each day that I live. And it's a beautiful ceremony.

Redhouse shares his music with many people from various walks of life, healing souls and reminding people of their core selves. Some are in grief. Others are in hospice awaiting their transition across the earthly line. Sometimes he plays his ethereal flute at the bedside of a dying patient. He does this, when invited, to "help them transition from this world to the next." He also works with the families, creating a peaceful, sacred space in which they can release their loved one. This is what he means by ceremony. He sees himself as an ambassador.

I'm looking in the doors and seeing the person lying there, and when I'm playing the flute, they wake up and they look at me. And then they go back in. I play very simply. It's not even words. It's not even a chant. It is a simple lullaby, a simple hum. And that's all that they want in that state when they are in transition between life and death. It is so comforting to them.

Sitting surrounded by his many and varied instruments, including bells, rattles, and drums, Redhouse shared with me a story of a woman who was on life support. He quietly watched her as he was playing his flute.

As I was doing that, I could feel myself going with her and experiencing what she was experiencing. I realized at that point that I have a lot of friends on the other side because for a lot of these people, when I come into the hospice the next time, their bed is empty. They're gone. But I have helped them during that transition. They have soared on my music, and they have gone into the next realm.

When working with people to transform their grief, Redhouse first uses the drum to create a pounding heartbeat. He then uses a second drum to create another heartbeat. And then he brings them together. He explained:

One heartbeat will represent the person that has passed on. The other heartbeat will be the heartbeat of the person that is doing the grieving. So these two heartbeats will be beating together.

Many times the grieving is because there's unfinished business, because there's something that was left undone, something that was not said that needs to be said. As I'm bringing the two heartbeats together and the two lives together, whether they're on the other side or here, they're still together. Their hearts are beating together. Then I will have that person that is grieving speak to the other person. And what they would have not been able to say, they will say to the other person.

## GRACEFUL PASSAGES

For Gary Malkin, founder of the multimedia publishing company Wisdom of the World, music and immersive media experiences offer a way to help people wake up to what matters most to them and help them remember who they really are. Malkin has had his own consciousness transformation that took him into the depths of art and living.

You live long enough, and enough losses come. The most significant one was not a literal death, but a death of my life as I knew it. I had been pursuing a life of fame and fortune as a composer for film and TV. I got very wrapped up in thinking that my value was my accoutrements—my house, my car, my wealth, my fame.

I won't tell the gory details, but I crashed. And everything died. Thank God, I still had my daughter, who was six at the time. Everything I had identified with had been taken away from me.

But then, I had been gifted with this wonderful idea: to help create a new kind of resource that would lubricate and catalyze and make easier the conversations about our mortality. It could be used at the key moment when people were facing dying in hospice circumstances and other environments where death was right around the corner.

Working with another musician, Malkin created an experiential program entitled *Graceful Passages*. Using what he called the last unregulated drug (music), the artists sought to expand people's capacity to accept what they previously saw as unacceptable. Music, combined with inspirational thoughts and images, has helped open people to an intimacy and authenticity around their views on living and dying. Malkin explained:

I call death the mother of all life. Everything that is in this life is defined against the relief of the awareness of the preciousness, the fragility, the breakability of this life that we are so fortunate to live. I thank my stars every day I'm alive. I know at any second it could end. That's what makes life so precious, when you really know. That's why I think this subject is so vital and important for human beings to engage with.

All the studies, all the religious traditions, verify that we are made of vibration and frequency and rhythm and tonality and resonance. This is what

we're made of, from the big bang on to now, in this
moment. I believe that the first sense to come into the
womb at twenty-four weeks is listening, and the last
to go is listening. To me, that's a little clue that says
there's something primary around sound that can link
up to the unseen world.

The power of sound, Malkin says, is affirmed by both his direct experiences and in the burgeoning science of sound. For the composer,
hearing with both our hearts and our minds can deepen our awareness around living and dying.

Music has a very important medicinal role for this
time, as a way to corral our intention into the present
moment. . . . It changes the air, and it brings you into
the present. That allows you to feel what's wanting
to be felt, which is the biggest part of the death
challenge—the denial, avoidance, and fear of what
it represents. We deny death because we all want
to avoid pain. Ironically, I've never felt more alive
and connected to my meaning than when I started
working on this issue. It's like the fountain of youth.
Facing your dying has a quality that's truly a portal
into radical gratitude for living. It's extraordinary.

## THE CRANE DANCE

Rumi, the thirteenth-century Islamic poet, is considered one of the
world's great mystic philosophers. His religious writings and poetry
are among the most beloved and respected in Islam and beyond.
After Rumi's death, his followers formed the brotherhood called the
Mevlevi, or whirling dervishes. In this brotherhood, dance is one of
the foremost transformative practices.

Elegant, smooth, graceful, the movements of the dance are
believed to connect the dancer to forces linking heaven and earth.
The practitioners move in a circular fashion, with one arm extended

toward the sky and the other toward the earth. They glide with a flow and purpose. Through the dance, they believe they are reborn in a mystical reunion with God. The dancers move along in unison, like a constellation of revolving forms. This dance is repeated over and over.

Bobaroğlu, the elderly Turkish imam whom we first met in chapter 6, explained the meaning of the dance as it relates to the practice of death awareness.

> In the Sufi tradition, after finishing the experience of visualizing one's own death, a dervish begins to connect to his master through an inner connection. And in this inner connection, the dervish must imagine his master's face in his dream and in his inner vision. He demands an inner connection with his master, heart to heart. The third degree of this inner connection is the imagination of himself in the presence of God.
>
> After these degrees of inner connection, the dervishes begin to realize some experiences. These are experiences of death itself, because they are without their body. When they go out of their body, and when they look at their body from above, they see a silver cord, like an umbilical cord, but a silver one, which comes from their body.
>
> We can also understand here the words of Jesus when he said, "When you are born from body, you are body. When you are born from spirit, you are spirit." It means, when we are born from our mother, we are bound with an umbilical cord with our mother. This is body to body, this connection. And then it is disconnected. The second cord is between the master and their disciple. This is the decision to devote themselves to someone. But the experience without a body means to be born from our own selves. That time, there is the silver cord. This is an inner connection. This is the religion.

## THE ART OF STORYTELLING

Music, movement, and voice offer means of expressing our worldviews. Words are both a reflection of our worldviews and a way of sharing these worldviews with others. Our narratives about death are embedded in our fears, hopes, and aspirations. For Jane Gignoux, a gifted storyteller and social artist, stories of death and dying help people imagine what may be coming in the death experience and beyond. Like others in this chapter, Gignoux believes that creating art can be a powerful practice for transforming our relationship to death. By engaging our imaginations, we may come to new insights or experience new sensations that expand our worldview. The storyteller explained:

> All forms of art that we know—from visual art, whether it's two-dimensional, three-dimensional; music; storytelling in poetry or theater; and kinesthetic art that we're moving through—are ways to express deep feelings and experiences that are hard to put into words.
>
> What happens is that when we see and hear, let's say, a beautiful piece of Mozart or Brahms, it touches us in a very deep way. It gets at maybe our gut—some deep, deep thing that we're hanging on to or that makes us feel more connected to a part of us that we hadn't been aware of. Art can help us release some of the pieces—maybe grief—that are holding on.
>
> And grief is natural. It's hard to accept all at once when somebody who's been very important in your life is no longer alive in a body. It takes time for us to go through the grieving. Art can help us be in touch with our own feelings. We can feel less isolated. Both looking at art and creating art, it doesn't matter.

## POETRY FROM THE SOUL

As Gignoux said, poetry and writing are powerful tools for expressing our connection to death and the departed. Setting intention, paying attention, and flowing with the experience of words can transform how we hold death. Elena Avila was born in El Paso, Texas. Attending Catholic school for her first eight years, she studied with the Sisters of Loretto, who, she noted, spanked her for speaking Spanish. She described her first Communion as a beautiful memory of herself in a little white dress, with her rosary and prayer book in hand.

Then she went to Mexico for the first time. It was there that she learned that her father had Aztec and Zapotec roots and her mother was Mayan and European. This mixing of blood and tradition led her to describe herself as a *mestiza*—a mixture of races that began interacting in the Americas over five hundred years ago, when the Spanish expanded into the New World and dominated the indigenous people. She found herself between cultures and worldviews when she learned about her history. When I spoke with her, she described herself as a "half-breed" who felt the oppression of growing up as a Chicana in the United States.

She shared some of the complexities of her own transformative journey: "It took a long time for me to find my place in spirituality. And then I realized that I believed in both, and that I could be Christian and I could be a spiritual indigenous person."

Avila ultimately became a Western-trained nurse. She also returned to Mexico to study with indigenous healers. She became an exemplar of an integrative healer, combining Western medicine and her folk practice of *curanderismo*. She explained that she would invite soul and spirit into her healing and nursing sessions. Also a gifted poet, she recognized the healing power of words.

At the time of my last conversation with her, she was living with advanced cancer and was very present in the moment. "This is what is right now," she noted, "and what are we going to do about that?" In her lyrical voice, she shared with me her own heartfelt message through a poem that she composed.

> Healing is remembering who we are—re-remembering,
> not being dismembered, but remembering.
>     We are already given to the power that rules our fate,
> and we cling to nothing so we have nothing to defend.
>     We have no thoughts, so that we can see.

We have no fear, so that we can re-member
ourselves, detached and at ease.

We will dart past the ego to be free.

I am a third-world woman god-worshiper, running
along the path of the sky, wearing my starry skirt and
tripping on myself as I make the sign of the cross.

Yes, I am different.

But there's enough space in the universe for
paradoxes, and I like being pyramidal, cruciform and
human, simultaneously.

I swish the ancient prayers in my mouth like holy gargle,
but I don't spit them out at you.

I rub your body with the sacred egg to help you
give your wounds to God.

Don't be chicken.

Let the egg be the instrument that picks up the
cosmic *caca* from your bodies.

Now, who did you think you are?

Superwomen?

Supermen?

I swallowed the myth, like consecrated wine,
because this earth is going to heaven when it dies.

You?

You can do whatever you want.

I am.

I just want to go to heaven on Earth's last sigh.

And I need your arms to fly. I need your arms to fly.

Avila passed away on March 17, 2011, in Albuquerque, New Mexico,
in the company of family and friends.

## THE SPACES IN BETWEEN

In the previous chapter, we met Fariba Bogzaran, a visual artist and
psychologist who has used art to navigate her own grief and to help
her with the departure of her beloved mentor, Gordon Onslow Ford.

Ford was an artist who painted every day. After his first stroke, and three weeks before he passed, he was unable to visit his studio anymore. After his second stroke, he was unable to walk. As she sat at the bedside of the ninety-one-year-old master artist, Bogzaran was amazed at how alert his mind remained. Three days before his death, she noticed that he started getting very anxious. Little by little, he began moving out of his embodiment.

During this time, Ford couldn't concentrate. This was very frustrating for him. Knowing that he was finding it increasingly difficult to meditate and therefore was getting more and more agitated, Bogzaran gave him a postcard of a painting of his.

> I said, "Gordon, just look at this painting. You've already painted where you're going." He looked up and says, "Brilliant." He just held it in his hand. Whenever he would get anxious, he would just pick the card and would look at his painting and would meditate inside of it.
>
> That painting is quite beautiful. It has all the different kind of what he called the "black hole" and the "white hole." It was the expression of the whole cosmos. And it was in some ways disembodied because you don't see any person inside of that painting. It was a vision of the whole inner universe for him. And here is a moment of him passing, and he can't go anywhere but into whatever he has painted. So it's almost like his painting was a precognition to his moment of dying, where he could have that as his medicine to go inside of his painting.
>
> And as soon as he would get anxious, he would be looking at the painting. And slowly, he went into coma and passed very peacefully.

Of course, not everyone is a gifted and talented artist. But is that important? Art can be a powerful tool for anyone. Art therapy uses art as a mode of expressing grief and other emotions. "The way one goes for the colors, and what colors attract people is important," Bogzaran told me. "For me, art is a mode of inquiry into consciousness."

Bogzaran also shared her most recent form of artistic expression: fringeing. She invited me into her studio. On a table in the middle of one room rested an ordinary piece of beige canvas. She had me sit down before it, and she guided me in a simple meditation that included working with the canvas. Very gently, I pulled one piece of thread at a time, gently deconstructing the fabric while creating something new and beautiful.

In a way, fringeing is an inversion of creation. Bogzaran reminded me to be patient and gentle as I pulled the threads, focusing on my breath in each moment as the threads fell away. She invited me to place my attention on the spaces between the threads and between each inhalation and exhalation of my breath. She encouraged me to think about the fringeing process as a life review. It is a powerful expressive practice, and the work of art is an important metaphor. As I mindfully pulled the threads, watching them release their places in the woven cloth, I imagined how we humans move through various life experiences before we are released from the cloth of life.

## MENDING BROKEN PLACES

Using art to help transform grief is powerful. When it engages a whole community, it is socially transformative. Lily Yeh had her own call to transformation. She was searching for meaning and authenticity in her life. She had been a studio artist with substantial success. One day, a little voice within her told her it was time to use her gift to help heal the deep suffering and grief that is so widespread on our planet. She stepped into the project and found the work so powerful that it changed the course of her life. Her method is to engage community members in collaborative art projects that bring beauty and solidarity to their neighborhoods.

Her transformative artwork has taken her to many places in the world. When I met with Yeh, she explained her efforts to mobilize what she calls the Barefoot Artists, Inc. Her goal is to inspire artists and change agents to develop community art projects to help inspire hope and mobilize social action. This has led her to collaborate with people from some of the most broken communities in the world, from an inner-city neighborhood in North Philadelphia to a barren churchyard in a vast garbage dumpsite community outside of Nairobi to a school for migrant children in China.

On one journey, Yeh went to Rwanda. Here, in 2004, she began the Rwanda Healing Project. Her goal was to help alleviate the deep despair that remained in the aftermath of mass genocide that took place in 1994. She found her way to a small survivors' village. Her artistic efforts were organized around a mass gravesite filled with skulls and bones of the murder victims. When she arrived, the skulls and bones were encased in a rough concrete rectangular mound with a rusted corrugated roof above it. "No one could image that beauty could exist in such a place," she expressed with passion. Depression and hopelessness were widespread among those who had survived. Using collaborative art, she engaged the community in the creation of a dynamic center to honor the dead and create a genocide memorial contained within a bone chamber. With little more than paint and broken pottery shards from which to create colorful mosaics, she enlisted numerous children and adults to help construct a healing space for the village. Her vision and her collaboration with the community transformed the bleak and depressing mass grave into a memorial that brings hope and dignity to people.

Yeh helped the survivors shift their experiences over their unspeakable loss by sharing their experiences, redefining their suffering, and using their broken lives to create places of beauty and solidarity.

"Piece by piece, people began to transform their grief," she explained. "In honoring their dead and bringing beauty to their dead, they begin to regain their human heart." Looking back, she was surprised that the project ever succeeded, but life beckoned and many people responded.

When I see brokenness, I also see the enormous
potential and readiness for transformation and rebirth.

We are creating an art form that comes from the heart and reflects the pain and sorrow of people's lives. It also expresses joy, beauty, and love. This process lays the foundation for building a genuine community in which people are nurtured by support and caring for each other.

In her own transformative process, as predicted in the worldview transformation model, Yeh has moved from "me" to "we." As her worldview has been transformed, she has helped shift the lives of many people in her wake. Her goal is not about art for art's sake, but to understand how creativity can be a fundamental tool for social healing. Through inspiration, education, and commitment, Yeh argues that art can help move individual grief into a force for living and growing. From the dark places on the planet emerge communities of hope where people can move from individual grief to collective resilience. Yeh's work reveals the ways in which creativity can help people to overcome their brokenness in order to reclaim their wholeness. In her eloquent words, "Creative action, guided by compassion, leads to transformation."

## GLEANINGS

Art allows us to engage multiple ways of knowing, which can open us to new insights about ourselves and our place in the larger network of relations. We can create art to process our emotions, engage our imagination, and live into dimensions of human experience that take us beyond our individual embodiment. Creating music, dancing, painting, storytelling, and writing can all serve as transformative practices, as they allow us to express our deepest and most authentic selves. The elements of transformative practice—intention, attention, repetition, guidance, and acceptance—all reveal the ways that art both reflects and transforms our fear of death. Art can be used for our personal transformation, helping us to find hope, while also speaking to the potentials of whole-systems change, one community at a time. Through beauty and hope comes the pathway to new beginnings.

# Revealing the Spaces Between

This practice, based on the lucid art of Fariba Bogzaran, invites us to experience the transformative practice of fringeing. You will need to find a piece of canvas. You can experiment with the best size for the cloth, but it helps to have something that is larger than two feet by two feet.

Place the canvas in front of you on a table. Very carefully, and with intention and attention, pull a thread out of the cloth. You will need to hold the canvas with one hand while you pull the thread with the other. Take your time, as rushing will lead to tangles. Watch the spaces between the threads reveal emptiness. Be aware of the moment when the thread finally breaks free from the cloth; feel its liberation from the whole. Put the threads in a pile as you continue fringeing, creating new artwork from that which has been transformed.

Repeat this act, staying mindful of the moment and the graceful way in which the thread moves through the cloth. Breathe into your feelings in the moment. Be aware of the space between breaths. Enjoy. You can decide how much fringe you want on your art. It is a wonderful practice to engage in over time.

After you have reached an appropriate stopping point, take ten minutes to record your impressions in your journal. Reflect on the way in which fringeing is a metaphor for life and death. Consider your own feelings and attitudes toward that space that lies between the threads, and the creation of beauty in the process of mindfully creating change in form.

# Chapter 10

# LIFE, DEATH, AND
# THE QUANTUM SOUL

We at least know that all of the consciousness
we have experienced in our lives cannot be
destroyed; it's stored somewhere.

RUDOLPH E. TANZI, PHD

Thomas Kuhn was a well-known historian of science. More precisely, he is a legend. He has helped dispel any illusion that our scientific models of reality are objective and absolute. His now-classic book *The Structure of Scientific Revolutions* explores the idea that our paradigms of reality are socially constructed.[1] Further, he showed that throughout history, paradigms shift periodically in ways that are revolutionary. Even in hard sciences like physics and chemistry, nothing is fixed.

Kuhn drew upon what is called the Copernican Revolution to illustrate a major paradigm shift. It was the Polish astronomer Copernicus who, during the Renaissance, discovered that the sun and not the earth is the center of our universe. His then-heretical discovery is today called the heliocentric model and predicts how the planets revolve around the sun, not around the earth. This radical idea led people of the time to rethink their worldviews, including their understanding of themselves, God, and the heavens. This catalytic worldview transformation ultimately led to the development of natural science and methods of empirical observation.

When paradigm shifts occur, according to Kuhn, new approaches and questions emerge. These shifts are disruptive because competing paradigms are often incommensurable and not easily reconciled. Skeptics and proponents of a worldview have always taken sides. In the Renaissance and at other times in history, those who challenged pervading paradigms were labeled heretics, banished from the prevailing religious institutions, and even burned at the stake. Today, scientists who challenge the dominant paradigm have been ostracized from their community of scientists or medical professionals, or they and their work have been the targets of smear campaigns. As we have seen in previous chapters, many transpersonal experiences have been reported by people of all ages and from many cultures. Still, science has dismissed such claims as fraud, delusion, or epiphenomena of the brain. Paradigm wars know no historical boundaries.

Paradigm shifts can also be moments of breakthrough, as we have witnessed throughout history. When a new worldview emerges that can be seen from many different vantage points, new possibilities open unexpected doors. Ideas that were previously dismissed are reconsidered in new light. In this chapter, we will explore some revolutionary ideas from the new field of post-materialist science that is offering fresh insights on age-old questions about life, death, and what may happen after.

## THE CONSCIOUSNESS REVOLUTION

It is undeniable that the billiard-ball logic of Newtonian physics continues to predict the physical world with great precision. We have a cloned cat named Carbon Copy, a chess champion named Deep Blue, and an orbiting international space station named *Discovery*. All provide testimony to the power of classical physics in controlling the world outside us. Any efforts to address the question of survival beyond death receive a simple answer: when you're dead, you're dead. That's it.

At the same time, reality is now being redefined as a complex quantum soup filled with nonlocality, probabilistic outcomes, string theory, cyberspace, cloud computing, biofields, potentialities in information fields, and even an uncertainty principle. Things are becoming faster and faster, smaller and smaller, and infinitely more complex and challenging to keep up with, even for the experts. In the midst of this shift, insights that bridge science and spirituality are now coming together to reveal new ways of understanding who we are and what we are capable of becoming. In particular, consciousness, once a strictly taboo topic in science, has become a spark that is shifting scientific, academic, religious, and social discourse.

Being present at the convergence of diverse and often conflicting definitions of reality offers mindboggling challenges. It also offers us an opportunity to reflect on our own worldview and to formulate—or reformulate—our understanding of life, death, and what may lie beyond that last breath. The worldview transformation model predicts that social transformation follows the same general pattern as individual transformation. It also suggests that the paths of both are more fractal than linear. Transformation can be messy. The breakthroughs that are emerging today are appearing at the intersections of worldviews, disciplines, and ways of knowing and being. The shift that is upon us represents a new ontology, or model of reality, beyond the senses and into expanded realms of being.

## CAN CONSCIOUSNESS SURVIVE THE BODY?

Does identity survive bodily death? And if so, how? Such questions define the discourse of post-materialist science. As we have seen in previous chapters, many cultures throughout history hold worldviews that when we die, something else lives on. How do these views match the physical model that defines Western science? In what ways does the post-materialist science of consciousness address noetic experiences of extended awareness, near-death experiences, communication with the deceased, and reincarnation? Bridging insights from both inner (noetic) and outer (rational) ways of knowing may well help us live into a new view of human possibility—now and after we die. As religion and science intersect, what emerges is an evidence-based spirituality that is an entirely new worldview for the twenty-first century.

Rudolph Tanzi, PhD, is one of the best spokespersons for this emerging worldview. His work combines impeccable scientific credentials with a deep spiritual practice. Tanzi is a professor of neurology at Harvard University. He also directs the Genetics and Aging Research Unit at Massachusetts General Hospital, where his research centers on the genetic causes of Alzheimer's disease. Tanzi also sees consciousness beyond the body. He is committed to establishing links between the seen realms of materialism and the unseen realms that lie beyond our physical embodiment. His views on consciousness and what may lie beyond bodily death are both unconventional and provocative. He explained to me:

> Scientifically we don't know if identity, self-awareness, can survive death. But one would think that in terms of developing a web of consciousness around yourself that interacts with all the consciousness in the universe, that's information. And information is the most basic thing in the universe.
>
> Information can be the structure of matter. It can be how energy is configured. We believe information cannot be destroyed. So we at least know that all of the consciousness we have experienced in our lives cannot be destroyed; it's stored somewhere.
>
> A neuroscientist will tell you that your identity is just within your neural network—that everything you do and learn is just associated with what you already know. So that leaves us with the question, when you die and the brain is gone, the electrical activity is turned off, is everything gone?
>
> The other side of the coin that most neuroscientists don't want to talk about is, where is consciousness? Where are memories? When you think about the past, where were they [the memories] stored? We don't have an answer for that in neuroscience. I ask students, I ask other professors this all the time. They say—it's all this hand-waving—"You know, it's in your neural

network." I'm like, "Where exactly?" "Oh, in the synapses." No, the synapses fire. [They fire] to recall the memory, but where's the actual memory? Where's my mother's face if I see it? What's the thumb drive of the brain that stores the JPEG of my mother's face? We have no idea.

Then the question becomes, is it stored here [pointing to his head], but not as a unified mass, which you would need for identity? Or is it actually coalescing as a global energy within a unified mass that we can call identity? And, for lack of a better word, we have the word *soul*. The soul is then the keeper of the identity. The consciousness you experienced over your life stays intact. I believe that. That's more of a spiritual belief right now than a scientific belief. But I trust my intuition more than anything, and my intuition says yes, this is probably the case.

For Tanzi, who we are is not defined by our physical experience. This may be surprising to hear from someone who has forged his career in mapping the molecules and mechanisms of awareness through our brains and bodies. In his unique way, he has found a worldview that offers an integration of what he knows from his scientific training and his own intuitive or noetic ways of knowing. Tanzi, like other post-materialist scientists, is suggesting that who we are transcends our brains and bodies. This idea points us toward new connections between personal experience and the soul as keeper of our identity. Death, within this worldview, is another phase in our ongoing transformation. This is a compelling idea that characterizes this twenty-first-century scientist and offers an emerging new paradigm for us all.

## CONSCIOUSNESS AND AN INTERCONNECTED UNIVERSE

Tanzi is not alone in his views of consciousness as a fundamental force in nature. Lothar Schäfer sees a similar vista of possibilities.

Schäfer is retired from the University of Arkansas, where he taught physical chemistry for forty-three years. Harbinger of a new worldview, Schäfer is optimistic that science is reaching a new way of understanding consciousness. Like Tanzi and other post-materialist scientists, he speaks about wholeness as the core of reality, countering the materialist and reductionist worldview that reduces the world to its parts. While he made his career in physical chemistry, measuring and manipulating the microscopic world, he sees the basis of matter as nonmaterial and the universe as interconnected.

"All things are connected," he explained to me. "Not in the empirical world, but in their nonempirical roots."

> The argument is this: if the universe is wholeness,
> everything comes out of it, everything belongs
> to it, including our consciousness. In that case,
> consciousness is a cosmic principle. The only chance
> you have that your consciousness survives when you
> die is that there is some consciousness outside. What
> may be in us is perhaps not our consciousness, but a
> cosmic consciousness.

In discussing his own cosmology, Schäfer acknowledges that a personal transformation linked his views of death with his scientific worldview. When he was younger, he was frightened about death. Today, he finds nothing frightening about it. Not that he has any clear opinion on what happens after. Still, as a post-materialist scientist, he grounds his own beliefs and assumptions in the meeting of science and spirit.

> In a way, there is not really a duality. That is kind of
> phrased in the mind-set of classical physics. I think
> there are different states of existence. Like when you
> leave a particle alone, it spontaneously goes over into
> a wave state of potentiality. This is why we can see
> interference patterns with double slits and electrons and
> so on. So there really is no duality at different states.

Take an ice cube. Put it into your drink. All of a sudden it's gone. That's what particles do, except they don't become something else material. No, they become a potentiality wave. That's what they do, and then if you do the right thing to it, the particle comes out. Where the mass was or went, I have no idea. I've asked a lot of physicists, and nobody has given me an answer to this.

If the universe is what classical physics says, what Newton thought . . . then the universe is a machine. It's nothing but particles running around following Newton's laws. It's closed because the state of the present determines the future. There can be nothing unexpected. It's like a clock. In a mechanical universe, people have a problem with the notion that our life is completely useless. We live at the edges of an alien world that doesn't care for our hopes, or for our pains, or for our crimes. It's a completely useless life. It's a life without dignity.

The only way we can have dignity is that the universe is not a machine. It's an organism like we are an organism. There is a cosmic mind with which we are connected. If there is a cosmic mind, it would be strange if it wasn't connected with ours.

## THE LEADING EDGE OF NOW

In this emerging post-materialist paradigm, there is a mutual inter-dependence between science and spirituality. There are no quick divides or easy points of separation. As Thomas Kuhn observed decades ago, while scientists are experts in charting the course of our physical nature, they carry within them their own metaphysics. Social psychologists have established that the values and beliefs of all of us, scientists included, inform our worldviews in key ways, below the threshold of conscious awareness. As the paradigm shifts, so do our metaphysical assumptions.

For Rick Hanson, cognitive neuroscience offers a trusted path for understanding our human potential. His writing and clinical work make use of a wealth of new data that points to the brain as the seat of experience. Through neurofeedback, we can reprogram our brains to reduce our suffering. Advanced brain mapping technologies are identifying neurochemical pathways that provide mechanisms of action for understanding our emotions, feelings, and beliefs.

And yet, reducing consciousness to brain states may not be enough to understand who we are and what may lie ahead after death. Research on meditation and contemplative practices such as prayer offers a glimpse into the further reaches of consciousness. Still, much of that research continues to be defined by the materialism paradigm in which extended states of consciousness are nothing more than correlates of brain states. For Hanson, like other postmaterialist scientists, it's confusing.

> I truly don't know for sure. My own personal experience and conviction is theistic; I think there really is a transcendental. My own version of that is that it's more an imminent transcendental view woven into the very nature of things for things to emerge. In other words, if the quantum physicists are right, and consciousness is required to enable quantum potentiality to coalesce into quantum actuality at the leading edge of now, at the emergent edge of now (which is, of course, the eternally emerging edge of now), if that's actually true, it would call for some kind of consciousness to be woven into the fabric of reality in an ultimate kind of sense for it to emerge and become solid, moment to moment.

Such a transcendental view offers an emergent model in which to consider the question of what happens to consciousness after bodily death. This model bridges our embodied nature with the complexities of a quantum universe. Further, it speaks to an evidence-based spirituality that grounds this emerging new paradigm.

## QUANTUM SOUL

To build on these complex and mind-boggling ideas, I sought further clarification from Stuart Hameroff, MD. An anesthesiologist, he is a professor and the director of the Center for Consciousness Studies at the University of Arizona. For many years, he has been seeking to understand the nature of consciousness and how it relates to the brain. His particular passion is the potential for quantum physics to explain the furthest reaches of consciousness.

In the high-tech surgical unit at the University of Arizona, Hameroff explained his own, somewhat controversial, views. In particular, he articulated a model he developed in collaboration with Sir Roger Penrose, the famed British mathematician. In Hameroff's view, which he hopes may explain the survival of consciousness after bodily death, microtubules are a key to consciousness. These protein structures are the body's scaffolding, out of which our skeletal system takes form. These structures are tubular and so, he says, can act as a kind of conductor for quantum events to occur in the body—and beyond. Like Hanson, Hameroff argues that death is not an on/off phenomenon. Instead, consciousness is seen as "the external edge of now."

> If consciousness is strictly an emergent property of complex computations, as most neuroscientists and philosophers agree, then there's no chance for an afterlife or consciousness after death. Its emergent property would end when the driving process ends. But if consciousness is related to quantum processes—for example, in microtubules, which we think are the quantum computers inside the neurons—consciousness and these quantum properties derive from the most basic level of the universe: space-time geometry. Consciousness, then, under normal

circumstances, is a process going on in our brains, in and around the microtubules inside the neurons. But deriving from the most basic level of the universe, the infinitesimally tiny Planck scale, it may occur between the ears, in the brain, in the microtubules.[2]

The Planck scale Hameroff refers to is a very small or a very large size in which quantum effects of gravity may occur. It is thought by quantum physicists that concepts like locality and causality break down at this scale. As such, the Planck scale speaks to a way in which consciousness may take on quantum qualities that transcend conventional views of space and time. During our interview, Hameroff continued:

> Under those circumstances, when the blood stops flowing and the brain starts to fail, the metabolic energy driving the quantum coherence is lost. But the quantum information, in space-time geometry, isn't necessary lost. And there's plenty of energy in the Plank scale (if you don't have to worry about the biology associated) for it to remain entangled as a unit. It can dissipate to the universe at large, remain entangled as a soul if you will, nonlocally, even perhaps something like holographically distributed, but still remain as one fundamental unit. I think it's plausible that consciousness can exist after bodily death, if it's related to space-time geometry, which I believe it is.

Hameroff then told me of a case study by an intensive-care specialist at George Washington University in Washington, DC. Hameroff explained that the palliative care physician Lakhmir Chawla was treating patients who were dying.[3] These patients and their families had elected to withdraw life support so that the patients would die peacefully. During this time, there were brain monitors on the patients. Hameroff described what happened:

When they started out, [the patients] had a level of physiological activity that was below what we would consider conscious in anesthesia. But there was some activity. As the heart stopped, and the blood stopped flowing, and the blood pressure pretty much went to zero, this number dwindled down to about zero, and the heart stopped. But then, in seven out of seven cases that Chawla studied, there was a sudden burst of brain activity that turned out to be gamma synchrony. This is a correlate of consciousness, that lasted anywhere from ninety seconds to twenty minutes, in one case.

Hameroff cited another study that was conducted at Virginia Mason University in Seattle.[4] In this study, three patients who were dying from brain injuries were taken into surgery to donate organs for transplants. The medical team monitoring the patients' brain activity as the patients died saw reactions similar to those reported by Chawla. Hameroff said:

They also found this burst of activity at the time of death, when the heart stopped. And Chawla suggested that this activity might correlate with what we know as the near-death experience, maybe even out-of-body experience. As you know, in many cultures for thousands of years, people have talked about the time of death. Sometimes spontaneously, the phenomena of the white light, being in a tunnel, visiting dead relatives, peace, tranquility, serenity, and this sense of calm are reported. Then they are revived, and they come back to life, and they are back to normal.

Of course, these patients [in both studies] died, and so we don't know. It doesn't say that they're actually leaving their body as in the out-of-body experience or that there's a soul leaving the body. But it does suggest that the near-death experience is something other than an effective hypoxia. Maybe it's something related to the near-death phenomenology of clarity, serenity, and so forth. I think it points in the direction that there could be something like an afterlife, although it certainly doesn't prove it.

Continuing to describe the science of consciousness, Hameroff articulated his view of a level of reality below atoms, where there is no mass. While cautious about his theory, Hameroff sees the implications of space-time geometry for understanding the mechanism by which reincarnation may occur.

There's some encoding of what mass becomes in the curvature of space-time. We think that consciousness, the precursors of consciousness at least, are irreducible, fundamental components of the universe. The way that you organize and put them together gives rise to the complex consciousness that we have.

When you come right down to it, space-time geometry is nonlocal. We know that from quantum theory and entanglement. Consciousness may be a phenomenon that repeats at different scales and is nonlocal and holographic, potentially. And it could be that this quantum information, in the case of reincarnation, goes back into another embryo or zygote, another set of microtubules which we think are the conveyers of this quantum information, or exist indefinitely at large in the universe.

Hameroff speculated on how his worldview might inform an emerging new worldview that bridges the inner world of spirit and the outer world of science.

If this turns out to be true, that consciousness can survive death, and I think it will to some extent, it's going to put science full circle and come back to the ancient traditions, which have been saying the same thing for thousands of years. Rather than a dualism of science and religion, there can be some synthesis and merging. I think that will be a good thing. I think it will be very gratifying to know that these possibilities exist. I don't think it will discount religion because, I think, living a good life will still be important. It will bring some kind of unity between science and religion.

## QUANTUM HOLOGRAPHY

Edgar Mitchell has been exploring the mystery of consciousness by bridging science and noetic insights. Mitchell is one of the Apollo 14 astronauts and a founder of the Institute of Noetic Sciences. He became a pilot at an early age and went on to fly for the US Navy in the Korean War. He then studied engineering at MIT, before becoming one of the few men to walk on the moon. When he landed safely back on Earth, he began asking big questions about life and the nature of reality.

For Mitchell, questions of death and a possible life after are fundamental to our understanding of reality. With new data coming from sources like the Hubble telescope, Mitchell shared that we are coming to a whole new understanding of the universe and what life is all about in the broadest sense. Like other post-materialist scientists, the former astronaut looks to quantum physics and holography to help explain concepts as enigmatic as reincarnation.

To help illuminate the connection, Mitchell first described how German scientist Max Planck, who won the 1918 Nobel Prize in Physics, discovered that all physical matter emits radiation. Some of that radiation has an associated charge—an electromagnetic area. Some of the radiation is just photonic, with no charge. Then in the 1990s, another German scientist, Walter Schempp, used complex mathematics to describe the radioactive emissions of every physical body photonically. He called his photonic description of the emissions a *hologram,* a three-dimensional image made from coherent light. It is a concept that has led to the development of modern medical technology, including the fMRI.

Working with Schempp and others, Mitchell has built a model that applies holographic information to our understanding of consciousness. He argues that there is a record of our thoughts and feelings in a holographic field. Like Tanzi, he asserts that there is a real record of stored information. Such a hypothesis may reveal a mechanism to explain concepts like nonlocal consciousness and reincarnation. Equating quantum holography with the ancient idea of Akashic Records, Mitchell argues that "nature doesn't lose its experience."

> That would mean that the experience of every life is preserved in the record somehow. And presumably it's recallable. . . . So if it is possible, for example, for a person to use the quantum holographic record of a prior person, to load it into the mind just the way you would load a computer program into a new computer, if that is true, that would thereby emulate that person's life and thought and reality.

He continued to elaborate on this theory to understand the nature of personal identity, acknowledging with appropriate caution that at this point it is still speculative.

> Our identity could be in some ways stored as photons in a quantum hologram. Is that an all-definitive statement? Not yet. Nevertheless, it is a correct statement. And exactly how much testing or what other testing might come out, or other laboratory experiments we might devise to either better authenticate or find the limitations on those applications, remain to be seen. We're right on the frontier here.

Ultimately, Mitchell sees the question of death and what lies beyond as informing how we live our lives. Pondering his own immortality, he expressed his own worldview about the survival of consciousness after bodily death and why it is important for how we live our lives.

> I think the more important thing for we humans is
> to learn to feel pleasurable, happy, successful in what
> we do in this life, and feel that we're being productive,
> caring, and helpful to each other and to our families.
> That that's really more important than whether we
> have all the answers to what happens after this life.
> Living this life to its fullest and properly and happily,
> to me, is far more fundamental.

## GLEANINGS

We are living at a time of enormous change. The new science of consciousness speaks to the emergence of a new paradigm that focuses on the powers and potentials of our minds. Post-materialist scientists have innovative ideas about the nature of reality—ideas informed by both their scientific training and their own spiritual beliefs and practices. Living into new discoveries that our identity may exist independent of space and time can give us a fresh new understanding of who we are—and what we may become beyond death. If we are, in fact, patterns of information, we may be both more and less than our personal identity.

This new worldview offers an expanded understanding of self that bridges our physical and our metaphysical beliefs. We are finding new ways of understanding life as part of a nonlocal, interconnected universe that exists outside of linear space and time. The reality that is emerging in the twenty-first century is dynamic, inclusive, and in some ways beyond words. In the next chapter we will see how, as this paradigm shift impacts science, it is also impacting society at large. In particular, a new worldview about death and the afterlife is impacting our system of healthcare, where death is often viewed as the enemy that must be conquered. As we transition to a new worldview that invites in the fullness of life, in all its brilliant facets, we may create a new vision for our human nature. And in doing so, we may find new paths to healing our cultural denial of death.

## Mindfulness in Nature

As we integrate the insights of this chapter, we may wish to reflect on the nature of our interconnectedness with all of life. In this practice, we will take time to be in nature. When you are ready, find yourself in nature. Take a walk in a park, the woods, or a garden. Be aware of the beauty around you. Focus your awareness on your body as it is embedded in your natural surroundings. Notice how your awareness moves along the path beneath your feet. Smell the air. Feel the breeze on your face. Stop by a tree. Reach out and touch the bark; experience the texture beneath your fingers, and feel your connection to the tree as a living system.

Close your eyes, take three deep breaths, and move your awareness inward. Sense the feelings in your mind and body as you connect to the tree. With each breath, be aware that you are growing more and more relaxed. Bring your attention to each breath, in and out. Be focused on that moment in each breath, experiencing only a sense of now in connection with another living system. There is no past or future, only this one precious moment, deep in nature, within your body, between your body and the tree, and within the breath that gives this moment life. Feel appreciation for the space within you that is fully in the present.

When you are ready, return your attention to where you are. As you return home, continue to hold this experience of being fully in the present moment and at one with nature with each step.

Afterward, take ten minutes to journal about your experience, exploring your own understanding of the nature of time and timelessness, independence and interconnectedness, inner and outer awareness. Continue to explore this topic over the days that follow, noticing ways in which you can be fully present in the moment and aware of your place in the natural world.

# Chapter 11

# HEALING SELF AND SOCIETY

Human behavior is not changed
through moral imperative, but through
a transformed imagination.

PAUL RICOEUR

As a physician and community leader, Karen Wyatt's purpose is to help society wake up from what she describes as a "cultural slumbering." Based in her years as a hospice doctor, she urges each of us to find the gift in every moment. In her work, she uses AWE as an acronym; it stands for "awake, willing, and engaged." By embracing our own potentials, we can willingly greet life with the same reverence as Wyatt's dying patients, but before we reach our final hours. As she explained, "We have to face our suffering in order to learn how to love as deeply as we need to love in this lifetime." Consistent with the worldview transformation model, Wyatt's personal transformation has led to a life that seeks to help catalyze a paradigm shift in both organized healthcare and in the lives of her patients.

Our healthcare system treats death as a scientific, moral, and financial tragedy, says Wyatt. It leaves those at the end of life in trouble. In the current medical model, death is seen as a failure. Wyatt, and increasing numbers of healthcare professionals, are seeking to change this perspective. Bringing AWE to healthcare can be a transformative practice that helps people at both individual and collective levels *to be*. It is a step of the worldview transformation model in action, using noetic insights about death to bring meaning and purpose to our lives and our organizations.

## EMBRACING THE HEROIC IMAGINATION

Our attempts to transform the fear of death take us back to the work of Ernest Becker, discussed in chapter 2. In his book *The Denial of Death,* Becker argued that our unconscious fear of death leads us to heroic efforts to defeat death. This "heroic imagination," in turn, leads to deep psychological and social pathology. Sam Kean, a student of Becker, explained it this way:

> With the Industrial Revolution, we became *homo faber,*
> makers of the world. The world was there for us to
> make and [in which] to construct meaning. Meaning
> wasn't given; it was something that we made. Death
> was not something that was seen as a part of a natural
> rhythm, a regenerative cycle of nature. It was seen
> more and more as an enemy that had to be defeated
> by our ingenuity—by our modern medicine especially.

Becker also discussed another kind of heroism—one that can help to transform our existential terror and bring about our individual and collective healing. This heroism involves becoming conscious pilgrims in the land of death. Breaking through our "character armor," the deep defense mechanism that denies death, allows us to confront our deepest fears. Kean continued, "All heroism occurs in the ambiance of the awareness of death or the repression of the realization of death." Facing the fear head-on can allow us to transform our worldview about life and how we engage it. It can, in Kean's language, prepare us to "participate in the meaningful direction of human history." Like Wyatt, Kean finds hope in the cultivation of wonder and awe.

> There's not an occasional miracle; the whole shooting
> match is a miracle. . . . And from all that flows
> gratitude and thankfulness and compassion. All those

virtues flow from an effort to live in the world in such a way that you're reverent of it. It's not esoteric.

## TRANSFORMING OUR IMAGES OF DEATH

Our cultural stories—and the images our society holds about death—shape the collective path we take forward. Willis Harman noticed this several decades ago. In a report called *Changing Images of Man,* Harman and his team of futurists at Stanford Research Institute articulated the paradigm shift we may be living into at this time.

> Images of humankind which are dominant in a culture are of fundamental importance because they underlie the ways in which the society shapes its institutions, educates its young, and goes about whatever it perceives its business to be. Changes in these images are of particular concern at the present time because our industrial society may be on the threshold of a transformation as profound as that which came to Europe when the Medieval Age gave way to the rise of science and the Industrial Revolution.[1]

Harman and his colleagues spoke of the dominant image that drives our social institutions and our efforts to control the forces of an objective world "out there," including our efforts to control our own mortality. If, in our shared mythology, defined by the biomedical model, we are victims of death, we may continue to suffer from our denial of death. Events like tsunamis, earthquakes, genocide, or terrorism shake our certainty; the crisis around death floods the shores of our social and economic worlds. We are being forced to examine our deepest assumptions about what is real and true.

In this process, new images are emerging to guide us. As we have seen in this book, the worldview transformation model predicts that crisis is a great catalyst for positive transformation. Even when we are in pain, we have the capacity to make shifts in our worldview and calibrate what gives us meaning and purpose. This holistic view places us within a living system, moving with the flow of evolution. We are collaborators, not dominators, in a new participatory model. Living into a new era of information, globalization, and quantum interconnections, the dominator image has begun to lose its focus, and the cultural discourse is shifting. Death is again being viewed as fundamental to the natural cycle of life.

## A TIPPING POINT

Our global society is at a tipping point; it's just not clear which way things are tipping. On the one hand, we may be on the verge of a full-systems collapse. We get daily reminders on the news about many ways in which we are in peril. Our collective fear of death is pushing us toward conflict and intolerance. On the other hand, if society moves forward as predicted in the worldview transformation model, we may be heading for the rebirth of a sustainable society. To find our way to a life-affirming option, we are well served to follow the advice of former writer and aikido master George Leonard: "Take the hit as a gift." Adversity is our opportunity.

An expanded awareness of death can enrich our lives. We are being called to heal a worldview that defines reality as nothing more than our physical nature. For Jean Watson, the key to transformation is healing relationships. As a nurse leader, she works to transform healthcare. Her goal is to transform human suffering into deep caring. She brings this awareness about the caring portion of healthcare to the way in which we treat death.

> We're engaged in helping to understand the difference between having pain with suffering and having pain without suffering. We're opening up an invitation for us to have a different meaning, or more meaning, of life purpose, another interpretation of death, and preparation for our own death, which ultimately leads to conscious dying as a possibility for us.
> I think we all hold a higher image and a higher vision of the other side that we haven't given ourselves

permission to engage in, explore, or to even have conversations around. And that's why people who are dying have so much to teach us. One person's humanity reflects on the other. So if we're shutting off that experience of the dying, we're shutting off our own experience of living. . . . There are opportunities for us as individuals, or as health professionals, or the public at large, to engage in these conversations as opportunities to ask new questions and find out so much more about what you or I are doing here.

Watson, like other visionary healthcare leaders, is advocating for a new model of medicine that sees death as a natural part of living. In this book, we have heard from people representing many of the world's traditions, spiritual and scientific. It is clear from their diverse voices that death need not be seen in extreme terms of crisis management. Making peace with death allows us to surrender into the natural cycle of life we are a part of. The shift in our view of death may include an end to the heroic measures that characterize modern medicine and end-of-life care. Each of the many people we met is calling for a model of death that is organic and free of fear. Through the convergence of diverse truth claims, we may chart our own spiritual wellbeing. Collaborating with nature, rather than exerting control over our own mortal being, allows us to harmonize with an evolving universe of possibilities.

## DEEPENING OUR COLLECTIVE CONVERSATION ABOUT DEATH

Conversations about death may be frightening, because death touches each of us at very unique personal and emotional layers of experience. Such conversations may also be deep and transformative.

Betsy MacGregor, MD, is especially sensitive to the complexities of how we think and talk about death. She worked in a busy city hospital for three decades, caring for seriously ill patients. In particular, she has focused on how her own profession relates to end-of-life issues.

"Until people in the healthcare professions really begin to look at their relationship to death and to dying, they don't bring themselves to caring for people fully who are facing that life situation," she told me.

Like Karen Wyatt, MacGregor uses AWE to help heal the healers who struggle to embrace their own emotional and spiritual needs. In turn, she hopes to improve the quality of care for dying patients. She urges her colleagues in organized medicine to open themselves to all their human dimensions.

> If we have closed off that part of our experience—
> maybe because we have had a difficult experience
> with a family member dying, or felt grief at the death
> of a patient, and have not really done the work of
> being with that experience, and allowing it to deepen
> in us and heal—then we have a closed door in us to
> caring for people who are approaching the end of
> their life.

MacGregor encourages healthcare practitioners, and all caregivers, to share their experiences and personal stories with others. When we share our stories, she says, we can cultivate a practice of deep listening that allows us to honor one another and those we care for and about.

> I believe something happens when health
> practitioners tell these stories. Talking about
> feelings is not something that people in our
> profession do easily. In fact, in our training, we're
> very often given the message that feelings don't
> count. We have to get feelings out of the way.
> They interfere with professional objectivity and
> our ability to make decisions clearly. . . . As a
> result, I think we keep a lot in ourselves that never
> integrates into our life experience, and it stays in a
> protected place.

Moving beyond such a protected place is not simple for any of us as we contemplate our own death or that of our loved ones. Nor is it easy to really ask ourselves questions about death and dying. The profound fear of death can limit our capacity to address these questions directly. For MacGregor, herself a breast cancer survivor, noetic insights can help us move beyond fear.

> I really think it's in our human nature to live with death. Death is not the opposite of life. It is part of life. If we're not acknowledging its reality, its presence with us at any moment, we're really selling ourselves short. We're closing ourselves off to a whole aspect of life. If we decide to begin to explore into that reality, I think there's something in us that will help us, because we have an inner knowing of who we are.
>
> Thinking about death and dying helps us begin to ask the questions: Who am I as a human being? How did I come to be here? Why am I here? What is this life that I have for? How do I want to use it? These questions all come along with the question of what is death? What is dying? And I think those are such fruitful questions. . . . People may experience fear in the beginning, but I think there's something in us that wants to ask them and is impoverished if we're not asking them.

While acknowledging the myriad challenges facing modern healthcare, MacGregor is optimistic. She sees a fundamental transformation in American culture regarding death and dying. She recalled the era twenty or thirty years ago when our relationship to birthing began to shift. At that time there was a move from labor and delivery being pathologized as a medical crisis to seeing them as a normal, joyful part of life. In the same way, death is becoming more normalized. MacGregor says:

> We're beginning to say in our culture, more and more, "I want to die at home. I don't want to die in

a hospital. I don't want my life prolonged further than I want it to be. I want the right to have a say in how I die." The right to die in the way people choose is something that's becoming acknowledged now in hospital settings, and it's beginning to change the way patient rights are seen in healthcare practice.

Bringing death back into our homes, witnessing it, sharing it, learning how to grieve with family and community—these are steps McGregor feels will change the stigma around death. She echoes Ernest Becker's views about the social implications.

The less we fear death, the less we will have other problems in our culture. I think that the materialistic greed in our culture is fed to a great degree by our fear of death, our denial of death, our effort to keep death from getting hold of us, and our . . . tendency to battle with each other. I think the more we see ourselves as having the same human experience of coming into life, living a life that has a purpose, and finishing our life . . . maybe that will chip away at that sense of separating ourselves from each other and identifying each other as enemies. I think it's all related, really.

## CONSCIOUSNESS IN ACTION

In this book, I have emphasized that death awareness can serve as a fundamental catalyst for our individual and collective transformation. We have the means to harness our shared intention and attention and use them to shift the widespread fear and terror associated with death. In this process, we can wake from a cultural trance that has estranged us from our natural relationship to mortality. Crafting a new story for humanity may help us to identify a more expanded set of possibilities for our lives and those of future generations.

As we saw in chapter 4, the indigenous people of the Ecuadorian Amazon dream for the collective. Perhaps, like these traditional people, we have a social responsibility to share our dreams. By doing so, we can begin to weave together the elements, the understandings, and the insights that come as we have these kind of soul journeys into the collective consciousness. Finding language to share our deepest callings may help us to create a common voice for individual and social healing around death.

Today a movement is afoot. There are people meeting in cafés, over dinners, in theaters, meeting rooms, and on the Internet to talk about death and its implications for how we live our lives. Through heartfelt and authentic conversations, people are beginning to change the shared discourse about death. There is healing that comes through conversation and the intimate sharing of stories and world-views. As we bring our own humility to the table and a spirit of not knowing into our boardrooms and to our bedsides, we may be led to expanded inner wisdom and shared insights. Our worldviews about death make a difference in how we live our lives. Post-materialist science and spirituality both invite us to see ourselves as part of an interconnected and mutually dependent web of life. Coming from a view of wholeness allows us to find our place in the natural order with grace and dignity. Finding the meeting point of noetic insights and rational knowing within us can lead to greater self-discovery and purposeful engagement in the world. As we change our views on life, death, and what may come after, we may better actualize our fullest human potentials—individually and collectively.

As Dean Ornish reminds us,

> we're all going to die. The mortality rate is still 100 percent. It's still one per person. To me, it's not how long we live, it's how well we live. When we embrace death, we can live life so much more fully and so much more joyfully because we realize we don't have all the time in the world. We can't say, "Well, I'll do that mañana." I want to embrace it and do it today whenever possible.

While there are no definitive answers about what happens after death, simply asking questions with an open mind and a warm heart opens our awareness to a new paradigm. It is a model of reality that is meta—big enough to embrace diverse views and perspectives. This new meta view allows us to think in new ways about how we may reinvent ourselves as a species. We may consider the ways in which our consciousness has limited us. We can ponder the ways that a fear-based paradigm has led us to maladaptive behaviors. Ultimately, we can begin to story together with a shared voice that offers hope and possibility. Living with an understanding of death is one of our birthrights. We can choose how we want to engage this fundamental part of who we are.

Taking a full-systems approach, we have the opportunity to see ourselves as part of an ecosystem that is defined not only by our existence here on planet Earth, but also by the role that we as humans play in an evolving story of the universe. As we shift our views on death, we are celebrating life. When we stop sweeping mortality under the carpet and treating it like a great taboo topic, we may face a new truth. Our model of reality becomes a relationship among relationships. And these relationships are based in consciousness. As Satish Kumar shared with me,

> relationship is intangible. It's nothing you
> measure. It's not something you can analyze.
> It's not something you can count or weigh. It's
> in consciousness. If you want to understand
> consciousness, you have to understand relationship.
> And when you understand relationship and
> consciousness together, then you realize that is the
> primary reality.
>
> Of this primary reality, it is love that lives on.
> Loving is eternal. Love never dies. Spirit never
> dies. It is eternal and also dynamic. It's evolving.
> It's unfolding. It's emerging. It's changing. Love
> expresses itself in different forms, in different ways,
> and at different times, but the essence, the quality
> of love, never dies.

## GLEANINGS

Each of us engages in a process of individual transformation. As we begin to share our views with one another, we can help shift our common vision to one that embraces the natural cycle of life and death. Breaking down our character armor may allow us to overcome our fear of death and live with greater purpose, balance, and harmony.

Recognizing death as a part of life may allow us to, as Sam Kean says, "participate in the meaningful direction of history." Bringing into conscious awareness the deep anxiety we share about death may lead us to a new paradigm. As each of us as individuals finds our own transformative path, so too may we revise the images in which our social institutions, particularly healthcare, treat death. We may shift our collective discourse from one of fear to one of awe and wonder. In this way, we may heal ourselves, our relationships, and the social world that we share.

◄ PRACTICE ►

## Imaging Your Loved One

Guided imagery can help you feel connected to your loved ones, both living and departed. Simple meditation practices can help support you during the grieving process and beyond.

Gently close your eyes and begin taking deep breaths. Start thinking of people you love, those who have brought meaning to your life. Begin to feel a sense of appreciation in your body. Feel it in your chest, your stomach, your back. Continue to breathe deeply, in and out.

Now think of a place that you love to be—a place that brings you peace. See yourself in that place.

Now, you are joined in that place by the person you have lost. Allow them to sit with you. Look at their face. In your mind, tell them how you feel about them. Feel the sense of connection to this person. Know in your body and in your mind that your love for this person never dies. It is simply expressed in different forms. Allow this feeling of love and connection to wash over you and to bring you a sense of deep peace. Know that you are part of an interconnected whole that has no boundaries and never ends.

When you are ready, thank your loved one for all that they have brought to your life. Bring your awareness back to the place where you are sitting. Reflect on your feelings, thoughts, and emotions. Breathe into a sense of peace and goodwill. When you are ready, take ten minutes to record your feelings in your journal.[2]

# CONCLUSION

Within this work, we've heard from remarkable people who have invited us to ask old questions in new ways. Who are we? What do we mean by *death?* What do we believe happens after? And why does this matter?

Engaging in these questions offers us many doorways for considering new possibilities. We have a fresh opportunity to think about our own mortality in an original way that expands beyond the boundaries of the dominant conversation in our time. By addressing the denial of death, we may liberate ourselves and our society from fear. We may spring forward with new forms of culture that support the fullness of our life (and death). Beyond maladaptive habits that deny death, there is a possibility that we can redefine the human enterprise as one of connection rather than separation. Through collaboration, we may begin to shift the social paradigm. We are alive in the midst of an enormous sea of change that can shake our world. Such change may be destabilizing, and it may also be transformative. We each have a role in shaping this future. Through our shared imagination, our transformative practices, and our conscious actions, we are forging a new model for living and for dying, as a seamless part of our shared metamorphosis.

As we seek to define this moment in human history, it's clear that we don't have answers to all these remarkable questions. It is possible that we can't even be sure what questions we should be asking. As we think about the opportunities that come from engaging our understanding around death and what possibly survives the body, core issues arise for each of us. With the convergence of the new post-materialist science and access to the world's religious and spiritual traditions, we are only beginning to formulate new understandings about our essence and our identity, who we are, and what we're capable of becoming.

Through my work and my life experiences, I have seen that the transformative process is a path worth taking. Changes in our worldview include shifts in both our inner and our outer realities. They link our direct personal experiences and our way of being in the world through action and service. They can lead us to a greater sense of joy, meaning, and love. Bringing our noetic understanding and our rational intellect together allows each of us to develop a deeper and richer sense of our connections to self, family, community, environment, and spirit. In this process, we may expand our awareness and appreciation for the sacred in every aspect of life. Social psychologist Daryl J. Bem explained it well when he said:

> The question is, can one alleviate the fear of death in some way? It doesn't have to mean that you have a belief in a particular afterlife—just being able to be at ease with what you have done and to have friends and relationships. One doesn't need what's thought of as a traditional view of the afterlife to be comforted by the fact that you may depart with friends who love you, that you've enjoyed life in some way or the other, and made the most of it.

In the face of fear-based images about death, we need an expanded sense of perspective, grounded in pragmatic hope. By creating new images that mark a new beginning, we may find the AWE-filled hero—an awake, willing, and engaged hero—within each of us. By harnessing our inner capacities—though self-reflection, meditation, contemplation, prayer, time in nature, direct and authentic conversations, and meaningful relationships—we can cultivate the resilience to navigate the challenges we face when considering our own and our loved ones' mortality. Out of catastrophe can come the renewal of civilization. Moving away from reactivity, fear, and panic toward emotional balance and positive collective actions allows us to apply the time-tested tools for sustaining our collective wellbeing, even in the face of death. In this process, we can promote deep healing—both individually and for our shared humanity.

A natural consciousness is readily available to each of us. It is not so much a supernatural awareness as it is an awareness of being present in this world. With awe and wonder, we can open ourselves to seen and unseen interrelations and interconnections with life. We can think in terms of patterns and relationships, rather than seeing ourselves as separate, isolated objects. As noted in a review article by social psychologists concerned with positive trajectories of terror management theory, "the dance with death can be a delicate but potentially elegant stride toward living the good life."[1] Let's shine up our dancing shoes and fully engage with one another in this precious performance of life. Death truly does make life possible.

# ACKNOWLEDGMENTS

This book was very much a team project. I have been supported by many wonderful people along the way—more than I can possibly list here. This work is the product of more than three decades of research on consciousness, transformation, and healing. Deep and often profound interviews with master teachers from the world's traditions, cultural healers, health professionals, and scientists representing many disciplines were collected along the way. In particular, I want to acknowledge the support and encouragement I received from the staff, board, and generous patrons of the Institute of Noetic Sciences, who offered me a safe place to explore audacious ideas for many years. Likewise, I want to thank the Center for Theory and Research at the Esalen Institute, where I was able to vivify my ideas about death awareness and the science of the afterlife in the context of beauty, conviviality, and a legacy of great thinkers who have inspired worldview transformation.

This book builds on a beautiful collaboration that led to publication of *Living Deeply: The Art and Science of Transformation in Everyday Life*. My teammates, including Cassandra Vieten, Tina Amorok, and Moira Killoran, helped me set the foundation for this work on transformation and death awareness by always showing up with their brilliance, courage, and tenacity. Several of the interviews I include here came from that phase of the work, and the worldview transformation model emerged from our hours of data analysis and creative exchanges. Likewise, thanks to Katia Peterson, who is a magnificent torchbearer for the worldview transformation project, filled with grace, aplomb, and boundless enthusiasm.

The companion to this book is a documentary film by the same title. Creating a feature film on life and death opened up many new directions in my work and helped shape my ideas for this book. I thank the Chopra Foundation for seeing the potentials of this project, and Deepak Chopra for encouraging me in many ways, including gifting me with the title *Death Makes Life Possible,* and for his mesmerizing foreword to this book.

I was supported in the interview process by the film's director, Mark Krigbaum, and the film's producer, Angela Murphy. We spent many creative hours together, sometimes losing sleep, in order to capture a complex topic through the lens of a camera. I also want to thank those who helped us record the interviews, including Mario Ayala, Phil Bissada, John Chater, Bill Cote, David Drewry, Kelly Durkin, Heidi Fuller, Michael Heumann, Joelle Jaffe, Brett Junvik, Martin Redfern, Dane Sawyer, José Vergelin, 4SP Films. Thanks also to Felicia Chavez and Davina Rubin for transcription. A very loud shout-out to Alan Pearce, who helped with many aspects of this project, always with cheer and awe-inspiring acumen. Charlene Farrell continues to be a source of support and commitment to quality. Jenny Mathews, my comrade in arms, coordinated the Kickstarter campaign that helped fund the recording of many interviews in this book (and the film). I also want to thank those generous people who invested in this work through the Kickstarter campaign. You are listed by name in the film and on the *Death Makes Life Possible* website (deathmakeslifepossible.com).

For the glossary, I relied on a variety of sources for writing the definitions. I want to thank those who helped me refine these definitions for accuracy, including Dick Bierman, Steve Braude, Larry Dossey, Brian Josephson, Stan Krippner, James Matlock, Vernon Neppe, Dean Radin, Charles Tart, and especially Catherine Poloynis and Chris H. Hardy. Any errors are my own.

Getting a book into publishable form is no simple feat. I want to thank my editor, Amy Rost, and the team at Sounds True. It has been a delightful process. I also want to acknowledge the folks at Speciality Studios, who helped me develop this project.

I would be remiss if I did not mention my beautiful family, who have helped me to become resilient in the face of loss and grateful in the face of laughter and delight. I include my best friend, Linda

Mendoza, who has taught me much about the blessings of life. I am thankful to my husband, Giovanni Mandala, who fills my world with music and chivalry, and who has helped me find the space to write this book. And to my beloved son, Skyler, who is my greatest teacher and source of hope for the future.

It is my belief that through conviviality and good will we may help transform the fear of death into an inspiration for living.

# INTERVIEW PARTICIPANTS

I am grateful to the wisdom holders who consented to be interviewed and whose insights provide the foundation for this book and for its companion documentary film, *Death Makes Life Possible.*

Aizenstat, Stephen, 2014

Alexander III, Eben, 2012

Artress, Lauren, 2012

Avila, Elena, 2009

Baker, Breese, 2012

Beckwith, Michael Bernard, 2014
(interview by Angela Murphy and Mark Krigbaum)

Beischel, Julie, 2012

Bem, Daryl J., 2011

Bobaroğlu, Metin, 2012
(interview by Michael Heumann)

Bogzaran, Fariba, 2012

Brinkley, Dannion, 2012

Chadly, Yassir, 2012

Chang-Lipsenthal, Kathy, 2013

Delorme, Arnaud, 2012

Fenwick, Peter, 2012

Gignoux, Jane, 2012

Greenberg, Jeff, 2014

Gu, Mingtong, 2011

Hameroff, Stuart, 2012

Hanson, Rick, 2011

Hufford, David, 2014

Jampolsky, Gerald, 2007
(interview by Cassandra Vieten and Tina Amorok)

Kawarim, Santiago, 2004

Kean, Sam, 2014

Kumar, Satish, 2012

Levine, Noah, 2005
(interview by Tina Amorok and Cassandra Vieten)

Lewis, Simon, 2012

Lipsenthal, Lee, 2011

MacAllister, Gloria, 2011 and 2013

MacGregor, Betsy, 2004

Malkin, Gary, 2012

Mathews, Jennifer, 2014

McMoneagle, Joseph, 2014

Mills, Paul, 2012

Mitchell, Edgar, 2012

Omer-Man, Jonathan, 2007
(interview by Tina Amorok and Cassandra Vieten)

Ornish, Dean, 2013

Pilcher, Josh, 2011

Radin, Dean, 2014

Rambo, Lewis, 2006
(interview by Cassandra Vieten)

Redhouse, Tony, 2012

Rousser, Margaret, 2012

Schäfer, Lothar, 2012

Sheldrake, Rupert, 2011

Shermer, Michael, 2012

Smith, Huston, 2006

Steindl-Rast, David, 2006

Tanzi, Rudolph, 2012

Teish, Luisah, 2011

Tucker, Jim B., 2012

Vieten, Cassandra, 2011

Walking Bull, Gilbert, 2006
(interview by Tina Amorok and Marilyn Schlitz)

Watson, Jean, 2011

Wyatt, Karen, 2014

Yeh, Lily, 2014

# NOTES

## Introduction

1. Marilyn Schlitz and William Braud, "Distant Intentionality and Healing: Assessing the Evidence," *Alternative Therapies in Health and Medicine* 3, no. 6 (1997): 62–73. Marilyn Schlitz, "Intentional Healing: Exploring the Extended Reaches of Consciousness," *Subtle Energies & Energy Medicine* 14, no. 1 (2003).

2. Marilyn Mandala Schlitz, Cassandra Vieten, and Tina Amorok, *Living Deeply: The Art and Science of Transformation in Everyday Life* (Oakland, CA: New Harbinger Publications, 2008).

3. Pew Research Center, "Daily Number: Baby Boomers Retire," December 29, 2010, pewresearch.org/daily-number/baby-boomers-retire/ (accessed February 7, 2014).

4. Marilyn Schlitz, "Nine Practices for Conscious Aging," *Spirituality and Health,* January 1, 2012. Marilyn Schlitz, Cassandra Vieten, and Kathleen Erickson-Freeman, "Conscious Aging and Worldview Transformation," *Journal of Transpersonal Psychology* 43, no. 2 (2011): 223–39.

5. California Healthcare Foundation, *Final Chapter: Californians' Attitudes and Experiences with Death and Dying,* February 2012, available as a PDF at the California Healthcare Foundation website chcf.org/publications/2012/02/final-chapter-death-dying (accessed February 7, 2014).

6. David C. Goodman, Elliott S. Fisher, C. Chang, N. E. Morden, J. O. Jacobson, Kimberly Murray, and Susan Miesfeldt, "Quality of End-of-Life Cancer Care for Medicare Beneficiaries: Regional and Hospital-Specific Analyses," *A Report of the Dartmouth Atlas Project* (2010).

7. Victoria Y. Yung, Anne M. Walling, Lillian Min, Neil S. Wenger, and David A. Ganz, "Documentation of Advance Care Planning for Community-Dwelling Elders," *Journal of Palliative Medicine* 13, no. 7 (2010): 861–67.

8. Barbara L. Kass-Bartelmes and Ronda Hughes, "Advance Care Planning: Preferences for Care at the End of Life," *Journal of Pain and Palliative Care Pharmacotherapy* 18, no. 1 (2004): 87–109.

9. Mount Sinai School of Medicine, "Health Care Spending in Last Five Years of Life Exceeds Total Assets for One Quarter of Medicare Population," press release from the Mount Sinai Hospital, February 10, 2012, mountsinai.org/about-us/newsroom/press-releases/health-care-spending-in-last-five-years-of-life-exceeds-total-assets-for-one-quarter-of-medicare-population (accessed February 8, 2014).

## Chapter 1: Transforming Our Worldviews

1. Pew Research Center Religion and Public Life Project, "U.S. Religious Landscape Survey: Summary of Key Findings," *Report 2: Religious Beliefs & Practices / Social & Political Views* (Washington, DC: Pew Research Center, 2008; available at religions.pewforum.org, "Full Reports" (accessed February 20, 2014).

2. George Bishop, "What Americans Really Believe," *Free Inquiry* 19, no. 3 (1999): 38–42.

3. Marilyn Mandala Schlitz, Cassandra Vieten, and Elizabeth M. Miller, "Worldview Transformation and the Development of Social Consciousness," *Journal of Consciousness Studies* 17, no. 7–8 (2010): 18–36.

4. Marilyn Mandala Schlitz, Cassandra Vieten, Elizabeth Miller, Ken Homer, Katia Petersen, and Kathleen Erickson-Freeman, "The Worldview Literacy Project: Exploring New Capacities for the 21st Century Student," *New Horizons for Learning* 9, no. 1 (2011).

5. Frances Vaughan, cited in Schlitz, Vieten, and Amorok, *Living Deeply.*

6. Rachel Naomi Remen, *My Grandfather's Blessings* (San Francisco: Riverhead, 2001).

7. William R. Miller and Janet C'de Baca, *Quantum Change: When Epiphanies and Sudden Insights Transform Ordinary Lives* (New York: Guilford Press, 2001).

8. Rhea A. White, "Dissociation, Narrative, and Exceptional Human Experience," in Stanley Krippner and Susan Marie Powers, eds., *Broken Images, Broken Selves: Dissociative Narratives in Clinical Practice* (Washington, DC: Brunner/ Mazel, 1997), 88–121.

9. Schlitz, Vieten, and Amorok, *Living Deeply*, 135–36.

10. Cassandra Vieten, Tina Amorok, and Marilyn Schlitz, "I to We: The Role of Consciousness Transformation in Compassion and Altruism," *Zygon* 41, no. 4 (December 2006): 915–31.

11. Adapted from Katia Petersen, Marilyn Schlitz, and Cassandra Vieten, "My Worldview," *Worldview Explorations Facilitator Guide* (Petaluma, CA: Institute of Noetic Sciences, 2012), 12.

## Chapter 2: Facing the Fear of Death

1. Ernest Becker, *The Denial of Death* (New York: Free Press, 1973).

2. Ibid., 87.

3. Abram Rosenblatt, Jeff Greenberg, Sheldon Solomon, Tom Pyszczynski, and Deborah Lyon, "Evidence for Terror Management Theory: I. The Effects of Mortality Salience on Reactions to Those Who Violate or Uphold Cultural Values," *Journal of Personality and Social Psychology* 57, no. 4 (1989): 681–90.

4. Jeff Greenberg, Sheldon Solomon, Tom Pyszczynski, Abram Rosenblatt, John Burling, Deborah Lyon, Linda Simon, and Elizabeth Pinel, "Why Do People Need Self-Esteem? Converging Evidence that Self-Esteem Serves an Anxiety-Buffering Function," *Journal of Personality and Social Psychology* 63, no. 6 (1992): 913–22.

5. Ibid.

6. Holly A. McGregor, Joel D. Lieberman, Jeff Greenberg, Sheldon Solomon, Jamie Arndt, Linda Simon, and Tom Pyszczynski, "Terror Management and Aggression: Evidence That Mortality Salience Motivates Aggression against Worldview-Threatening Others," *Journal of Personality and Social Psychology* 74, no. 3 (1998): 590–605.

7. Tom Pyszczynski, Abdolhossein Abdollahi, Sheldon Solomon, Jeff Greenberg, Florette Cohen, and David Weise, "Mortality Salience, Martyrdom, and Military Might: The Great Satan Versus the Axis of Evil," *Personality and Social Psychology Bulletin* 32, no. 4 (2006): 525–37.

8. Ibid.

9. Becker, xiii.

10. Kenneth E. Vail, Jacob Juhl, Jamie Arndt, Matthew Vess, Clay Routledge, and Bastiaan T. Rutjens, "When Death Is Good for Life: Considering the Positive Trajectories of Terror Management," *Personality and Social Psychology Review* 16, no. 4 (2012): 303–29.

11. Vieten, Amorok, and Schlitz, "I to We."

12. "Death Makes Life Possible: Mapping Worldviews on the Afterlife," online telecourse, January 23, 2013. Marilyn Schlitz, Jonathan Schooler, Alan Pierce, Angela Murphy, and Arnaud Delorme, "Gaining Perspective on Death," *Journal of Spirituality and Clinical Practice* 1, no. 3 (2014): 169–80.

13. "What is Attitudinal Healing?" The Hawai'i Center for Attitudinal Healing, ahhawaii.org (accessed September 26, 2014).

## Chapter 3: Glimpses Beyond Death and the Physical World

1. Joseph McMoneagle, *The Stargate Chronicles: Memoirs of a Psychic Spy* (Charlottesville, VA: Hampton Roads Publishing Company, 2002).

## Chapter 4: Cosmologies of Life, Death, and Beyond

1. Diana Eck, "The Age of Pluralism," Gifford Lecture Series from University of Edinburgh, January 2009.

## Chapter 5: Science of the Afterlife

1. Dean Mobbs and Caroline Watt, "There Is Nothing Paranormal About Near-Death Experiences: How Neuroscience Can Explain Seeing Bright Lights, Meeting the Dead, or Being Convinced You Are One of Them," *Trends in Cognitive Sciences* 15, no. 10 (2011): 447–49.
2. Dirk De Ridder, Koen Van Laere, Patrick Dupont, Tomas Menovsky, and Paul Van de Heyning, "Visualizing Out-of-Body Experience in the Brain," *New England Journal of Medicine* 357, no. 18 (2007): 1829–33.
3. Daryl J. Bem, "Feeling the Future: Experimental Evidence for Anomalous Retroactive Influences on Cognition and Affect," *Journal of Personality and Social Psychology* 100, no. 3 (2011): 407–25.
4. Ian Stevenson, *Reincarnation and Biology: A Contribution to the Etiology of Birthmarks and Birth Defects* (Westport, CT: Praeger, 1997).

## Chapter 6: The Practice of Dying

1. David Phillips, Gwendolyn E. Barker, and Kimberly M. Brewer, "Christmas and New Year as Risk Factors for Death," *Social Science & Medicine* 71, no. 8 (2010): 1463–71.
2. Christine Evans, James Chalmers, Simon Capewell, Adam Redpath, Alan Finlayson, James Boyd, Jill Pell, John McMurray, Kate Macintyre, and Lesley Graham, "'I Don't Like Mondays'—Day of the Week of Coronary Heart Disease Deaths in Scotland: Study of Routinely Collected Data," *The BMJ* 320, no. 7229 (2000): 218–19.

## Chapter 7: Grief as a Doorway to Transformation

1. Elisabeth Kübler-Ross, *On Death and Dying* (New York: Macmillan, 1969).

2. George A. Bonanno, *The Other Side of Sadness: What the New Science of Bereavement Tells Us about Life after Loss* (New York: Basic Books, 2009).

3. John Schneider, "The Transformative Power of Grief," *Institute of Noetic Sciences* 12 (1989): 26–31.

4. Karen M. Wyatt, *What Really Matters: 7 Lessons for Living from the Stories of the Dying* (New York: SelectBooks, 2011).

5. Lauren Artress, *Walking a Sacred Path: Rediscovering the Labyrinth as a Spiritual Tool* (New York: Riverhead Books, 1995).

## Chapter 8: Dreaming and the Transformation of Death

1. William James, *Principles of Psychology,* vol. 2 (London: MacMillan and Co.,1891), 296.

## Chapter 10: Life, Death, and the Quantum Soul

1. Thomas S. Kuhn, *The Structure of Scientific Revolutions,* 2nd ed. (Chicago: University of Chicago Press, 1970).

2. Stuart Hameroff and Roger Penrose, "Orchestrated Reduction of Quantum Coherence in Brain Microtubules: A Model for Consciousness," *Mathematics and Computers in Simulation* 40, no. 3 (1996): 453–80.

3. Lakhmir S. Chawla, Seth Akst, Christopher Junker, Barbara Jacobs, and Michael G. Seneff, "Surges of Electroencephalogram Activity at the Time of Death: A Case Series," *Journal of Palliative Medicine* 12, no. 12 (2009): 1095–100.

4. David B. Auyong, Stephen M. Klein, Tong J. Gan, Anthony M. Roche, DaiWai Olson, and Ashraf S. Habib, "Processed Electroencephalogram during Donation after Cardiac Death," *Anesthesia & Analgesia* 110, no. 5 (2010): 1428–32.

## Chapter 11: Healing Self and Society

1. O. W. Markley and Willis Harman, eds., *Changing Images of Man: Prepared by the Center for the Study of*

*Social Policy/SRI International* (Oxford, UK: Pergamon Press, 1982).

2. The "Imaging Your Loved One" practice is inspired by Lee Lipsenthal in *Death Makes Life Possible* (the film), 2013.

## Conclusion

1. Kenneth E. Vail III, Jacob Juhl, Jamie Arndt, Matthew Vess, Clay Routledge, and Bastiaan T. Rutjens, "When Death is Good for Life: Considering the Positive Trajectories of Terror Management," *Personality and Social Psychology Review* 16, no. 4 (2012): 303–29.

# GLOSSARY

**advanced multichannel EEG**   An EEG "channel" is a voltage signal represented by a difference in voltages between two electrodes. It is used to measure changes in the brain's electrical activity.

**Akashic Records**   From the Hindu concept of *akasha,* the belief in an immaterial field of potentials that is encoded with the information about all beings and things.

**animistic worldview**   Embraces nonhuman entities (animals, plants, mountains, weather) as possessing a spiritual essence. The word comes from the Latin *anima,* or "soul."

**anxiety-buffer hypothesis**   Asserts that strengthening self-esteem should reduce anxiety in response to awareness of one's mortality. Formulated by social psychologists, including Jeff Greenberg, based on Ernest Becker's work on the denial of death.

**Attitudinal Healing International**   This global network, founded by Jerry Jampolsky, promotes healing to help remove self-imposed blocks such as judgment, blame, shame, and self-condemnation that limit people's ability to experience lasting love, peace, and happiness.

**AWE**   An acronym used by hospice physician Karen Wyatt to reflect her approach to being "Awake, Willing, and Engaged" with dying patients.

*ayahuasca*   A psychedelic drink made out of *Banisteriopsis caapi* vine, either alone or with the leaves of shrubs from the genus Psychotria. It's used for reconnecting with one's deep Self and the spirit world, as well as for divinatory and healing purposes by Amazonian native peoples.

**consciousness**   Process that involves our awareness of ourselves and the world.

*curanderismo*   This Latin American folk-healing method includes the *yerbero* (herbalist), the *partera* (midwife), and the *sobador* (masseur). Some US regional healthcare plans cover services of *curanderos.*

**death-thought accessibility hypothesis**   Predicts that if individuals avoid thoughts about death by investing in a worldview that buffers their self-esteem, then when threatened, they should have more death-related thoughts than when not threatened. Is an extension of the terror management hypothesis of Jeff Greenberg and colleagues.

**depth psychology** (also called Jungian psychology)   Psychoanalytic techniques by Carl Gustav Jung that posit a subject to the personal unconscious (the Self) having a drive for spiritual self-realization and to a "collective unconscious" as a deep interconnection between all human beings and the planet.

**Dream Tending**   A dream-work method that considers dream images as living images in order to gain insights into the unconscious. Developed by Stephen Aizenstat, Dream Tending is based in depth psychology.

**empiricism**   A theory that knowledge comes from sensory experience and is the basis of experimental science.

**epistemology**   The means by which we know what we know.

**esoteric**   Knowledge or insights that are intended to be understood by a small or exclusive group of people.

**evidence-based spirituality**   A form of spirituality in which scientific methods and findings are used to understand and explain questions of a religious nature, including the mysteries of life and the evolving nature of the universe.

**evolutionary neuropsychology**   This transdisciplinary approach brings together social and natural sciences to examine psychological traits such as memory, perception, and language from a modern evolutionary perspective.

**exoteric**   Knowledge or ways of communicating intended to be understood by the general public.

**functional magnetic resonance imaging (fMRI)**   A functional neuroimaging procedure that uses MRI to measure brain activity by detecting associated changes in blood flow.

**hero myth**   Joseph Campbell defined a classic sequence of actions that are found in many stories and myths that share three stages: Departure, Initiation, and Return. Campbell presented this theory in his book *The Hero with a Thousand Faces*.

**Heroic Imagination Project**  Founded by social psychologist Philip Zimbardo to teach people how to embrace their heroic impulses in order to take effective action in challenging situations.

**hologram**  A three-dimensional image formed by the interference of light beams from a laser or other coherent light source.

**hot-sauce paradigm**  Various amounts of spicy hot sauce were used to measure aggression in terror management theory experiments conducted in 1999 by Jeff Greenberg and associates. The method demonstrated a participant's aggression toward a target, without inflicting actual harm on the person eating the hot sauce.

**humanistic healthcare**  An interdisciplinary field that seeks to bring humanistic values and principles into organized healthcare, including open communication, mutual respect, and emotional connection between healthcare professionals and patients.

**in-group bias**  A pattern of favoring members of one's social group with whom they share values over "out-group" members who may represent a different value system or cultural background.

*ip'ori* (from Yoruba Lucumi tradition)  That part of a person that is connected to spirit, that always has been and always will be.

**Jainism**  An Indian religion that prescribes nonviolence and emphasizes spiritual equality between all forms of life.

**Labyrinth Movement**  A labyrinth is a circuitous path that is walked for meditative purposes. Labyrinths have been used for thousands of years worldwide. The modern Labyrinth Movement began in 1990 when Rev. Lauren Artress, Canon of Grace Episcopal Cathedral in San Francisco, published *Walking the Sacred Path*.

**locus coeruleus–noradrenaline system**  The principal site for brain synthesis of norepinephrine, a chemical released in response to stress. This brain region and the areas of the body affected by the stress hormones are collectively described as the locus coeruleus-noradrenergic system.

**magnetic resonance imaging (MRI)**  A technology that uses a magnetic field and radio waves to create detailed images of the body's organs and tissues.

**magnetoencephalography**  A noninvasive technique that detects and records the magnetic field associated with electrical activity in the brain.

**Mahayana Buddhism**   One of the three main existing branches of Buddhism, specifically that of Tibet and Northern India. The path of the bodhisattva seeking complete enlightenment for the benefit of all sentient beings is referred to as "Mahāyāna."

**microtubules**   Components of the cytoskeleton, found throughout the cytoplasm, that function on a binary basis. They have been hypothesized by Stuart Hameroff and Roger Penrose to have an immense capacity for a digital-type processing at a subcellular level that may help insulate quantum events in the body.

**mortality salience**   According to the terror management theory, when human beings contemplate mortality and their vulnerability to death, feelings of terror emerge unless they are brought to conscious awareness under circumstances that bolster self-esteem.

**Naqshbandi**   A major spiritual order of Sunni Islam Sufism that traces its spiritual roots to the Islamic prophet Muhammad.

**near-death experience**   A widely reported personal experience associated with clinical physical death that often includes the impression of being outside one's physical body, a vision of deceased relatives or religious figures, a sense of peace and wellbeing, a tunnel of light, and a feeling of unconditional love and acceptance.

**neuroimaging**   The use of various technologies to image the structure of the nervous system, including both structural and functional dimensions.

**neuronet**   Networks of living neurons in the brain.

**neuropsychiatry**   A branch of medicine that focuses on mental disorders caused by diseases of the nervous system.

**New Thought–Ageless Wisdom**   A tradition of spirituality grounded in the principle of being a compassionate, beneficial presence on the planet that's taught by Michael Bernard Beckwith, founder of the Agape International Spiritual Center.

**noetic experience**   A state of insight that involves feelings of "inner knowing" that are not associated with rational thought or discursive intellect and that have transformative potentials for the individual.

**nonlocality**   In quantum physics, a correlation and inter-influence between particles that transcend the constraints of 4-D space, time, and local causality.

**nonlocal consciousness**  Refers to aspects of our consciousness and processes that appear to transcend the constraints of space and time.

*olam ha-ba*  The Hebrew concept of the afterlife or World to Come.

**ontology**  A system of knowledge that organizes what we understand to be true or real.

**parapsychology**  The scientific study of psi phenomena.

**Planck scale**  The scale (in cosmology and physics) at which the Planck constant is effective. It implies the smallest quantity of energy (a quantum) and time (of the order of magnitude of 10 power minus 44 of the first second of our universe), at which particles, space and time (thus matter) can now exist. The Planck scale is an immense time before the Big Bang or inflation phase (at about 10-36 to 10-34 second).

**positron emission tomography (PET)**  A nuclear medicine imaging technique that produces a three-dimensional image of functional processes in the body.

**post-materialism**  The transformation of individual values from materialist, physical, and economic to values that embrace humanistic and transpersonal qualities, multiple ways of knowing, interconnectedness, and whole-systems thinking.

**psi**  A term used to describe telepathy, clairvoyance, precognition, and psychokinesis. The focus of parapsychology research.

**psychophysiology**  The branch of psychology that focuses on the physiological aspects of psychological processes.

*qi*  A central underlying principle in traditional Chinese medicine and martial arts. *Qi* is frequently translated as "life force" or "energy flow."

**qigong**  A traditional Chinese medicine practice of aligning body, breath, and mind with qi for health and wellbeing.

**quantum entanglement**  A quantum mechanical phenomenon in which the quantum states of two or more objects have to be described with reference to each other, even though the individual objects may be spatially separated. As a result, measurements performed on one system seem to be instantaneously influencing other systems entangled with it (as proven through EPR-paradox experiments). Extended to signify any process functioning beyond space and time constraints.

**quantum hologram**  In "The Quantum Hologram and the Nature of Consciousness," Edgar D. Mitchell and Robert Staretz propose a new model of information processing in nature that seeks to explain how living organisms know and use information in ways that involve quantum properties of nonlocality.

**quantum processes**   In quantum mechanics, a quantum process describes any event and behavior of waves-particles at the quantum scale (that is the wave-particle scale).

**radical empiricism**   A philosophical doctrine put forth by William James that argues that any realm of human experience is a valid domain for science, an argument that is at odds with scientific materialism, which focuses only on the physical dimensions of reality.

**scientific materialism**   The Western scientific framework that defines reality by its material nature and holds that mental phenomena are epiphenomena or byproducts of the brain that derive from neurophysiological processes.

**social philosophy**   The study of ethical questions and values about social behavior and interpretations of society and its institutions.

**social psychology**   The scientific study of how people's behaviors, thoughts, and feelings are impacted by the presence of others—whether actual, implied, or imagined.

**space-time**   1. The merging of the three dimensions of space and one dimension of time in Einstein's relativity theory. 2. The geometry and/or fabric of space-time in different theories, e.g. viewed by Einstein as curved and storing energy in itself.

**super-psi hypothesis**   Predicts that the results of communications with the deceased that involve ostensibly accurate information (such as a spirit reading with a medium) may reflect telepathy among the living as compared to communication with the deceased.

**telomere**   A compound sequence of nucleotides at the end of a chromosome that's associated with stress and aging.

**terror management theory**   Hypothesizes that the terror of mortality motivates much of human behavior and that by understanding and mitigating this terror, humans can avoid destructive behaviors. Based in the work of Ernest Becker, this theory was developed in social psychology by Jeff Greenberg and colleagues.

**Theravada Buddhism**   The oldest surviving branch of Buddhism that means "the Teaching of the Elders."

**transformative practices**   Formal and informal practices that involve a shift in perspective or worldview including three components: psychological (changes in understanding of the self), convictional (revision of belief systems), and behavioral (changes in lifestyle).

**transpersonal psychology**   A school of psychology, started by Abraham Maslow and colleagues, that integrates the spiritual and transcendent aspects of human experience.

**worldview**   Frameworks of ideas, beliefs, and perspectives about ourselves, others, and the world that are held by individuals and groups of people and that are largely unconscious.

**Worldview Explorations Project**   An educational program based on research of worldview literacy and transformation by Marilyn Schlitz and colleagues to increase awareness of worldviews through the Institute of Noetic Sciences.

**worldview transformation model**   Predicts that transformations in worldview begin with noetic experiences that instigate a process of exploration, transformative practice, and beneficial changes in perception, behavior, and health. It's based in research summarized in *Living Deeply: The Art and Science of Transformation in Everyday Life* by Schlitz, Vieten, and Amorok.

**Yoruba Lucumi**   A traditional animist cult (of Southwest Africa) based on "possession trance" (as opposed to shamanic trance) that birthed Caribbean voodoo and some Brazilian cults. In colonial and slavery times, Yoruba Lucumi integrated features of Catholicism.

# INDEX

Achuarian traditions, 71–73
Agape International Spiritual Center, 36, 87
agnosticism, 90–91
Aizenstar, Stephen, 154–57
Alexander, Eben, 58–60
Allen, Woody, 107
Amorok, Tina, xxi
anxiety-buffer hypothesis, 24, 25
art, 167–80
    dance, 172–73
    music, 167–72
    storytelling, 174
    visual arts, 176–78
Artress, Lauren, 84–86, 123–24, 137–39, 151, 162–63
atheism, 91–93
attention (tranformative element), 16
Avila, Elena, 175–76
AWE (awake, willing, and engaged), 199, 204
ayahuasca, 72

baby boomers, xxi-xxii
Baker, Breese, 145–46
the Barnum effect, 25
Becker, Ernest, 23, 26, 200
Beckwith, Michael Bernard, 36–37, 87–88
Beischel, Julie, 107, 142
Bem, Daryl J., xxii, 22, 110–11, 212
the Bhagavad Gita, xi

Bishop, George, 3
Bobaroğlu, Metin, 125–27, 173
Bogzaran, Fariba, 157–59, 176, 181
Bonanno, George, 133–34
the brain, 104–6
Brinkley, Dannion, 57–58
Buddhism, 5, 33

C'de Baca, Janet, 14
Chadley, Yassir, 60–63, 67, 78–80
change, embracing of, 7–10
    see also transformation
Christian traditions, 82–90
community, 39–40
consciousness, ix-xi, 184–89, 191–95
cycle of life, 4, 5, 9, 21–22, 77–78

dance, 172–73
death
    Achuarian traditions, 71–73
    and animals, 4, 136
    and art, 167–80
    as awakening, 87–88
    and the body, 22, 27
    change, embracing of, 7–10
    Christian traditions, 82–90
    cycle of life, 4, 5, 9, 21–22, 77–78
    denial of, 23–27
    and dreams, 151–65
    economics of, xxiii

embracing, 30–33
end-of-life decisions, xxii-xxiii
fear of, 21–41, 206
as a gift, ix, xii
a "good death," 122–34
holidays honoring, 144–48
and identity, 34–36
Islamic traditions, 60–63, 67,
   78–80, 125–27, 172–73
mystery of, 41
Native American traditions, 50,
   67–69, 73–74, 168–70
preparing for, 125–29
resistance to discussing, xxii-xxiii,
   203–6
science of, 98–119
timing of, 124–25
visualizing, 125–27
death-thought accessibility hypothesis,
   24, 27
Delorme, Arnaund, 107–9, 112
*The Denial of Death* (Becker), 23, 26,
   200
Día de los Muertos, 147–48
dreams, 151–65
   Dream Tending, 154–57
   lucid dreaming, 158–59
drum rituals, 168–70

Eck, Diana, 94
EEGLAB, 107
emotions, 128–29
empiricism, 47
end-of-life decisions, xxii-xxiii, 203–4
eternity, 79

fear of death, 21–41, 206
Fenwick, Peter, 97, 98–103
Ford, Gordon Onslow, 176–78
fringeing (art therapy practice), 178,
   181

Gallup poll, life after death, 3
Gignoux, Jane, 174
*Glimpses of Death* (film), 99
God, xvi-xvii, 36
Graceful Passages, 171
Great Spirit, 68, 69
Greenberg, Jeff, 23
grief, 132–49
   phases of, 133–34
   practices for transforming, 137–42
   rituals (shared grief practices),
      142–48, 168–70
guidance (tranformative element), 16
Gu, Mingtong, 8, 75–77

Hameroff, Stuart, 191–95
Hanson, Rick, 4–6, 118–19, 190
Harman, Wllis, 201
healthcare practitioners, 204–5
heaven, 83, 86
hell, 83
heoric imagination, 200
hero myth, 26
the hoop dance, 73–74
holography, 195–96
hot-sauce paradigm, 25–26
Hufford, David, 160–61

identity, 34–36
Institute of Noetic Sciences, 11, 56,
   108, 128, 195
intention (tranformative element), 15
ip'ori, 75
Islamic traditions, 60–63, 67, 78–80,
   125–27, 172–73

James, William, 43, 98, 102, 153
Jampolsky, Jerry, xxvi, 39–40
Jewish traditions, 80–82
judgment, 29
Judgment Day, 86

Ka, 68
Kawarim, Santiago, 71–73
Kean, Sam, 26, 27, 90–91, 200
Krishna, xi
Kübler-Ross, Elizabeth, 133
Kuhn, Thomas, 183, 189
Kumar, Satish, 34–36, 77–78,
    121–22, 208

Laburinth Movement and practice,
    137–39, 149–50
laughter, 139–40
Levine, Noah, 33–34
Lewis, Simon, 43–47
life after death, 3
life cycle. See [/ital] cycle of life.
limbo, 82
Lipsenthal, Lee, 1–3, 22, 38, 124
    Chang-Lipsenthal, Kathy, 153–54
Living Deeply: The Art and Science
    of Transformation in Everyday Life
    (Schlitz, Vieten and Amorok), xxi,
    11, 16–17
loneliness, 39–40
love, 38–39
lucid dreaming, 158–59

MacAllister, Gloria, 148
MacGregor, Betsy, 203–6
Malkin, Gary, 167, 170–72
Maslow, Abraham, 37
matter, 88–90
"matter first," xv
Matthews, Jennifer, 139–40
McMonealge, Joseph, 51–55
meditation, 31–32, 129
mediums and mediumship, 106–12,
    141–42
Melami, 125
memory, x-xi, xiii
Memorial Day, 144–47
Miller, William, 14
the mind, 101

"mind first," xv
Mitchell, Edgar, 195–97
Mobbs, Dean, 105
mortality salience hypothesis, 24,
    25–26, 27
MRI scans, 104–5
music, 167–72
Muslim traditions. See Islamic
    traditions.
My Grandfather's Blessings (Remen), 12

Naqshbandi, 126
Native American traditions, 50,
    67–69, 73–74, 168–70
natural cycle. See cycle of life.
near-death experiences, xix, xx, 43–44,
    51–60, 98–103
    see also noetic experiences
neuroscience, 104–6
noetic experiences, 43, 44–47, 48,
    49, 60
    see also near death experiences

objective versus subjective, 107, 120
the observer effect, xi
Omer-Man, (Rabbi) Jonathan, 80–82
On Death and Dying (Kübler-Ross),
    133
Ornish, Dean, 30–32, 207
The Other Side of Sadness (Bonanno),
    133–34
out-group, 27
out-of-body experiences. See near-
    death experiences

paradigm shifts, 183–84
past-life experiences, 1–3
PET scans, 104–5
Planck, Max, 195
Planck scale, 192
pluralism, 93–94
post-materialist paradigm, 189–90
practices

acceptance and self-compassion, 130

connecting to the world, 65–66

dream recall, 164–65

fringeing (art therapy practice), 181

imaging your loved one, 209–10

labyrinth practice, 149–50

"me" to "we," 96

nature, mindfulness of, 198

self-esteem, 42

subjective versus objective ways of knowing, 120

worldview exploration, 18–19

the present, being in, 37–38

purpose, 33

qi, 75

qualia, xii

quantum changes, 14

quantum holography, 195–96

quantum soul, 191–95

quantum theory, xi, xiii

the quantum field, xi

Quinn, Janet, 123

Radin, Dean, 56–57

Rambo, Lewis, 82–84

Redhouse, Tony, 21, 38, 73–74, 127–28, 163–64, 168–70

reincarnation, 34, 112–19

    see also past-life experiences

Remen, Rachel, 12

repetition (tranformative element), 16

Reynolds, Pam, 56–57

Ricouer, Paul, 199

rituals (shared grief practices), 142–48, 168–70

Rousser, Margaret, 4, 136

Rumi, 172

Rwanda, 179

sanskara, xi

Schäfer, Lothar, 187–89

Schempp, Walter, 195–96

Schlitz, Marilyn

    dream life, 151–52

    *Living Deeply: The Art and Science of Transformation in Everyday Life* (with Vieten and Amorok), xxi, 11, 16–17

    near-death experiences, xix, xx

Schneider, John, 134

the self, xv

Sheldrake, Rupert, 70–71, 161–62

Shermer, Michael, 63–64, 91–93

*Skeptic* magazine, 63

Smith, Huston, 88–90

the soul, 85, 191–95

spiritualism, 141–42

    *see also* mediums and mediumship

Steindl-Rast, David, 8–9, 37–38

Stevenson, Ian, 112–13

storytelling, 174

*The Structure of Scientific Revolutions* (Kuhn), 183

subjective versus objective, 107, 120

Sufism, 125–27, 172–73

suicide, 30–33

suffering, 33

super psi, 110

Swami Satchidananda, 31

Tanzi, Rudolph E., 183, 186–87

Teish, Luisah, 21–22, 74–75, 131–33, 141, 144, 163

terror management theory (TMT), 23–27, 64, 70, 156

transformation, 7–17

    change, embracing of, 7–10

    five key elements, 15–16

    noetic experiences, 44–47

    transformational model, 13

*Trends in Cognitive Science* (journal), 105

Tucker, Jim, 113–18
Twain, Mark, 21

Vail, Kenneth, 27
Vaughan, Frances, 10
Vieten, Cassandra, xxi, 121, 128–29
visual arts, 176–78

Wakan and Wakan Tanka, 68
wakes, 144
*Walking a Sacred Path: Rediscovering the Labyrinth as a Spiritual Process* (Artress), 137
Walking Bull, Gilbert, 67-69
Watson, Jean, 48–50, 202–3
Watt, Caroline, 105
Wellspring Institute for Neuroscience and Contemplative Wisdom, 4
*What Really Matters: 7 Lessons for Living from the Stories of Dying* (Wyatt), 136
whirling dervishes, 172
White, Rhea, 14
wisdom, 134–36
Wisdom of the World (publishing company), 170
worldviews, 3–19, 22
    change, embracing of, 7–10
    definition of, 6
    and fear of death, 30
    transformational model, 13, 32, 35, 47, 128
    transforming, 10–17
    *see also* transformation
worldview transformational model, 13, 32, 35, 47, 128
Wyatt, Karen, 134–36, 199

yanifa, 74
Yeh, Lily, 178–80
yoga, 139–40
Yoruba traditions, 21–22, 74–75, 141, 144

# ABOUT THE AUTHOR

Marilyn Schlitz is an award-winning author, charismatic public speaker, pioneering researcher, and leading voice in the emerging fields of consciousness studies and mind-body health. Marilyn received her PhD in social anthropology at the University of Texas, Austin, and was awarded a post-doctoral fellowship in social psychology at Stanford University. She served as president and CEO at the Institute of Noetic Sciences (IONS). She's the founder of Worldview Enterprises, where she creates cultural media and education programs on consciousness, healing, and worldview transformation.

A prolific writer, Marilyn has published hundreds of articles in peer-reviewed journals and popular magazines, including former columns in *Spirituality and Health Magazine, The Noetic Review,* and *Shift* magazine. She serves on the editorial boards of *Explore* and the *Permanente Journal.* Her books include *Consciousness and Healing: Integral Approaches to Mind-Body Medicine* (with Tina Amorok and Marc Micozzi) and *Living Deeply: The Art and Science of Transformation in Everyday Life* (with Cassandra Vieten and Tina Amorok). Her acclaimed documentary film (with Deepak Chopra) entitled *Death Makes Life Possible* serves as a companion to this book.

Marilyn is presently a senior fellow at IONS, senior scientist at the California Pacific Medical Center Research Institute, fellow of the Academy of Transdisciplinary Learning and Advanced Studies (ATLAS), and serves on the board of directors for Pacifica Graduate Institute. She lectures widely throughout the world and in various venues. Her scientific findings have been featured in the *New York Times, Huffington Post,* and on NPR, PBS, and elsewhere. Marilyn lives in Northern California with her husband, teenage son, and two dogs. To learn more about her work, blog, and public speaking, see her website at marilynschlitz.com.

# ABOUT SOUNDS TRUE

Sounds True is a multimedia publisher whose mission is to inspire and support personal transformation and spiritual awakening. Founded in 1985 and located in Boulder, Colorado, we work with many of the leading spiritual teachers, thinkers, healers, and visionary artists of our time. We strive with every title to preserve the essential "living wisdom" of the author or artist. It is our goal to create products that not only provide information to a reader or listener, but that also embody the quality of a wisdom transmission.

For those seeking genuine transformation, Sounds True is your trusted partner. At SoundsTrue.com you will find a wealth of free resources to support your journey, including exclusive weekly audio interviews, free downloads, interactive learning tools, and other special savings on all our titles.

To learn more, please visit SoundsTrue.com/freegifts or call us toll free at 800-333-9185.

SOUNDS TRUE
many voices, one journey